CASEBOOK:
ALIEN IMPLANTS

by Dr. Roger K. Leir, D.P.M.

WITH AN INTRODUCTION BY Whitley Strieber

A DELL BOOK

Published by
Dell Publishing
a division of Random House, Inc.
1540 Broadway
New York, New York 10036

PHOTO CREDITS: Photographs and artwork provided courtesy of Dr. Roger Leir and MUFON Ventura–Santa Barbara.

Dell books may be purchased for business or promotional use or for special sales. For information please write to: Special Markets Department, Random House, Inc., 1540 Broadway, New York, N.Y. 10036.

Dell® is a registered trademark of Random House, Inc., and the colophon is a trademark of Random House, Inc.

ISBN: 0-440-23641-X

This book was first published by Granite Publishing, LLC, with the support of the Bigelow Foundation in cooperation with the National Institute for Discovery Science.

First published as Volume IV, The New Millennium Library.
Published by arrangement with Granite Publishing, LLC.

Printed in the United States of America

Published simultaneously in Canada

August 2000

10 9 8 7 6 5 4 3 2 1

OPM

D E D I C A T I O N

To my beloved parents, Shirley and Joseph,
who have stood by me and guided me in this writing,
as if they still lived,
and
to the individuals who became my surgical patients.
Their lives touched mine in most unusual ways.

Whitley Strieber brings to light what others hope to conceal in *Whitley Strieber's Hidden Agendas*. One of today's most respected names in the field of paranormal research, Strieber gathers together the best available evidence on such controversial issues as conspiracy theories, UFOs, close encounters, and other unexplained phenomena. Drawing on his vast knowledge and experience, spanning the globe and exploring a wide variety of subjects, Strieber brings us unprecedented access to information from reliable sources—revelations that will rock our deepest beliefs and may open a door to worlds other than our own.

FROM THE FRONT LINES OF MODERN MEDICINE . . . REAL-LIFE STORIES OF ALIEN IMPLANTS AS TOLD BY THE DOCTOR WHO REMOVED THEM!

✧ A mother of three has been aware of two small metal objects in her toe ever since the night she had a terrifying close encounter of the third kind. . . .

✧ Best-selling author Whitley Strieber became haunted by what he knew was an implant in his ear. While he was undergoing surgery to remove it, something extraordinary occurred. . . .

✧ A man has a strange, unexplainable metallic object in his thumb. Upon extraction, it proves identical to the mysterious implants removed from other victims. . . .

The evidence mounts, in . . .

CASEBOOK: ALIEN IMPLANTS

A C K N O W L E D G M E N T S

Writing this book has been an occasion for me to reflect gratefully on some people who have profoundly influenced my life and, consequently, this book.

The first and foremost was my dear father, Joseph Leir. When I was a child, he impressed me with his general knowledge and understanding of the world. I wondered if he derived it from his experience as a 32nd-Degree Mason and Shriner or whether he had an uncanny ability to read between the lines and see the truth.

My cousin, Dr. Kenneth Ring, also influenced me through our conversations and through his book, *The Omega Project,* which is concerned with the relationship between near-death experiences and UFO close encounters. I extend my heartfelt appreciation to Ken for his advice and friendship.

My wife, Sharon, though not directly involved in the work reported here, has been of enormous moral support to me, and I am deeply thankful to have her beside me in life.

I am fortunate to be surrounded by the wonderful people of the Ventura–Santa Barbara, California, Mutual UFO (MUFON) organization. They assisted me in organizing and preparing for the surgical procedures described here. One of the most creative and thoughtful of them is our state section director, Mrs. Alice Leavy. This fine lady encouraged me to perform in areas far beyond what I had originally considered my field of expertise. Her encouragement has been pivotal to the course I am now seeking.

My thanks also go to:

Whitley Strieber and his wife, Anne, who inspired in me the courage to pursue the scientific study of objects that are not only foreign to the human organism but also to our planet itself.

Raymond Fowler, one of the most outstanding researchers in the field of ufology, who took an interest in my work and gave me the honor of contributing to his book, *The Andreasson Legacy*.

Robert O. Dean and Cecelia Dean. They gave me encouragement to continue my abduction research. They pointed me in the right direction and made me aware of the importance of this line of research.

My dear friend, Dr. Tal, who stayed at my side and helped me with his medical and surgical skills, guiding me through troubled times of financial despair. Because of Dr. Tal's many abilities, I and my associates were able to proceed with the research reported here.

CASEBOOK:
ALIEN IMPLANTS

I N T R O D U C T I O N
By Whitley Strieber

The African veldt, high noon. In the distance, wildebeest graze. Closer, a pride of lions lounge beneath an
acacia. They lie in the slack postures of digestion, looking almost dead as flies swarm around them. None
watch, none take notice of the four human beings about
a quarter of a mile away.

Why would they? The people are downwind from
the lions, the lions are asleep. But it wouldn't matter
even if they were awake. In the absence of scent, the
sight of the nearly stationary figures would have so little
meaning for them that they would only gaze curiously.

In fact, this happens. The most ferocious of the

lionesses awakens. Her eyes scan the shimmering horizon. She sees the sticklike figures. Their slow movements arrest her attention. But not for long—they have little meaning for her, and her gaze is soon directed elsewhere.

One of the men has a gun, but the indifferent lions don't know its significance. Nothing in the world can enable them to understand the danger of a gun. They will be surprised by its sound, but they will never relate the disaster that befalls one of them to the noise.

The dart gun makes a thud loud enough to awaken the lions. The wildebeest, also made uneasy by the noise, surge and whistle. Crows rise squalling from the tree that shades the lions. For her part, the female feels a sudden burst of pain, sharp enough to make her roar in anger. In another moment, though, she finds herself in blackness.

There is nothing, not the slightest trace of awareness, nothing at all.

The men come closer, moving warily, expertly. They drive off the other lions with shots and shouting, and soon the female they have stunned lies alone on the ground, her eyes open, her great mouth drooling. As they draw close, they wave at the legions of flies. The other lions, now disturbed, watch nervously. It is highly unlikely that they will attack, but the men take no chances. The intrusion into the life of the lions is handled like a squad-level military operation. Four of the men take up positions between the female and the other lions. The fifth works quickly, approaching the lion, grasping her heavy head in his thickly gloved hands, turning it, then performing a brief surgery in the loose skin around her neck.

Blood spurts, then stops as he quickly makes a couple of stitches. He steps back, listens to a small hand-held radio. As soon as he's sure the beeping is steady, he turns it off and puts it in one of the many large pockets on his coverall. At his signal, the men fade away from the pride, and life among the lions goes back to normal.

Or, almost. The female now bearing the implant has no direct memory of what happened to her. But the fact of anesthesia does not block all memory. She is disturbed, uneasy; her neck hurts. Sometimes, when the object now nestling behind her right ear touches the bone, she hears the beeping. This makes her roar and pace—and the object moves and the sound stops and she settles down.

It takes weeks for her to come back to normal, weeks and weeks. But the wound heals, the pain eases, the inner turmoil subsides. In the end, she ceases to notice the occasional beeping.

None of the lions can even begin to conceive that she and her whole pride are being tracked every moment of their lives from a laboratory in Kansas City, Kansas. The world of the trackers—the satellite that picks up the transmissions, the computers that make sense of the data, the young biologist who uses it—it's all completely inconceivable to them. No amount of knowledge could change this. The brain of the lion simply cannot process the information.

Now shift to another setting. It's the dead of a May night in upstate New York. The year is 1994. Deep in

the country, there is a cabin in which two people are
sleeping, a husband and wife.

Into the silence there comes a car. It's a white Jeep
Cherokee. It pauses at the gate, but only for a moment.
The automatic lock is opened by a gloved hand that
comes out of the driver-side window, deftly punching
the correct keys. At exactly the right instant, the car
proceeds. Its speed is governed so carefully that it will
reach its destination within a hundredth of a second of a
predetermined time.

Inside the car, a thin woman in a khaki coverall
opens a small, very finely made black case. She
removes a needle, then draws a small amount of liquid
into it.

The bedroom overlooks the gravel driveway; the
windows are open. The man sleeping inside is more
capable than the lioness. He has a better memory and
better defenses. In fact, he has been so distressed by a
series of events like this that he suffers from
post-traumatic stress disorder. One of his symptoms is
guarded sleep, which means that he sleeps more like a
wild animal than a human being, rarely descending into
the deep sleep that most of us enjoy.

He awakens as soon as he hears the faint crunch of
gravel caused by the tires of the approaching vehicle.

Even so, it's too late for him, far too late. The
approach of the car was among the last events in an
elaborate operation. Half an hour before, another man
in another car had come down another road behind the
house; a third man has been standing outside the house
with a small device made by the E-Systems division of
Raytheon. It is an extremely sensitive microphone,

tuned to receive the slight sonic signatures of human breathing and movement. An onboard computer chip interprets the information and rates the mental and physical status of the individual. The computer can filter out any other sounds.

As this man waits, his receiver trained on the sleeping couple inside, two people leave the car. Training a high-intensity magnetic field generator on the garage, they freeze the burglar alarm's trip switches. With a duplicate garage-door opener, they enter the house.

Inside, the intended victim is now fully conscious. He's heard the crunching of gravel on the driveway below his window. He knows that something is very wrong.

Normally, a target's being this fully awake would stop the operation. But it can't stop now. With what is going on above the house, nobody involved would dare stop.

From behind the house, the team's observer calls out, "Condition red." The penetration group moves more quickly.

Following a familiar path, they hurry upstairs and into the bedroom. There is another figure behind them, gliding along. The intruders do not look behind them, but they move as if they are well aware that the figure is there. They reach the bedside of their target, the figure following close behind. The man in the bed, now wide awake, starts to resist. With their magnetic field generator on another setting, they immobilize him. The small figure now comes around in front of them and leans over the man, closing his eyes with its fingers.

The two human beings probably have little or no

idea what is being done to this man. As it is done, they stand silently by, watching. The instant the figure is finished, there is a powerful flash of light from above. Everybody races from the scene, and silence again enfolds the night.

This would be science fiction, except that it is a description of events that happened to me on the night of May 24, 1994. As I was awake most of the time, I remember events quite clearly. I do not know if I have described the instruments used correctly, but I would submit that my experience was nothing more than another version of what happened to the lioness on the veldt, and what happens to animals every day as scientists implant them with tracking devices for research purposes.

Late the next afternoon, I noticed that there was an object embedded in the pina, or stiff outer surface, of my left ear. I had already discovered that the garage door was wide open, even though the alarm system had remained operative. I remembered being awakened during the night by the sound of gravel in the drive, then seeing the people and the shadowy figure enter my bedroom. I remembered the faces of the people quite clearly—the woman with her dark eyes and pretty, delicate features, and the man behind her with his distinctive beard. As for the third party—I only saw that shadow.

At the time, I was deeply disturbed by what had happened. Again and again, my fingers returned to the object in my ear. Sometimes it would make a strange sound, a sort of rough burring noise. My ear would get hot and turn bright red at those moments. I became more electrically sensitive than I had been before. It became routine for streetlights to go out when I drove or walked under them, and electrical appliances would turn on or off when I entered a room.

I was not like the lioness. I knew and understood that the object was there. I had complex, conflicted emotions about it. Did I want it or not? Was it evil or not? Was it dangerous?

I did not know where to turn to have it examined. I couldn't tell my doctor what had happened—there was at that time nobody studying alien implants.

Then, in 1996, I heard of Dr. Leir. By that time, the object had been observed during one of its periods of activity by William Mallow, a staff scientist at a prominent research institute in Texas. Tests had been done to see if a signal could be acquired from the object, but they had been unsuccessful.

Then came August of 1996. Roger Leir and his team of surgeons were going to remove implants from a group of people, and witnesses were being invited to watch the surgeries on closed-circuit television. My wife, Anne, and I went, and I walked unsuspecting into one of the most moving experiences of my life. Tears rolled down my face as I watched these gentle, brave people submitting themselves to their surgeries. I could hardly even say why I was crying. But the event so completely overwhelmed me that I had to leave, just to get out for a little while, just to get some air.

In the pages of this book, Dr. Roger Leir will tell you of how it felt to be the surgeon responsible for removing these objects. He will offer overwhelmingly powerful evidence that the objects could not have entered the bodies of the witnesses by chance. He will reveal laboratory data that will convincingly suggest that the objects are artifacts of a higher technology.

And he will write of his own persecution, and the efforts to silence him.

The motivation to ignore the existence of these objects—and this is a constant theme among scientists and intellectuals, who certainly ought to know better—is emotional, not rational. However, because government remains silent and science refuses to face the truth, our society offers people who have these objects embedded in them no help.

Are the objects dangerous? What do they do? How are they affecting our lives?

For somebody who actually bears an implant, these are agonizing questions. I sat there in that surgery wondering, "What about me? What about the thing in my ear?" By this time I was extremely pessimistic about my chances of getting it out. In part, this was because it was obviously very active. But there was also another reason.

A few years before, a witness named Betty Ruth Dagenais had waited until after death to have an object from the pina of her ear autopsied. This was because she believed that she'd been told that it would kill her to have it removed during her lifetime. Her object had been similar to those being removed by Dr. Leir and his colleagues: a tiny piece of dark metal. In her case, an engineer who examined it thought that it might have characteristics something like those of a radio crystal that would be capable of sending and receiving signals.

I began to get compulsive about feeling the object in my ear. Sometimes it seemed a little more prominent, other times less. I thought of the people who'd put it in, remembered the story of the lioness. When I gazed up at night into the ocean of stars, I wondered if somebody up there might be monitoring me. Just as the movements of the big cat were being constantly downloaded from the loneliness of the veldt to a cluttered laboratory in Kansas, maybe things about me were being delivered to somebody much farther away, for reasons that I could hardly even begin to guess.

The sense of being watched was eerie, and I worried that maybe my mind itself was being controlled. But when Dr. Leir offered his services to me, I remembered the warning that had prompted Betty Dagenais to wait until after her death to have the object removed, and I was scared.

Unlike the lioness, who would have soon put her ordeal behind her, I could not stop worrying about my implant. Sometimes, during the intimate silence of meditation, it would turn on, its rough screeching shattering my concentration. I would feel the ear getting hotter and hotter, and I would wonder what was happening to me and all the other close encounter witnesses—what was *really* happening?

Finally, in May of 1998, I decided that I could not stand to have the implant in my ear any longer. But I did not want to have my surgery done in California. I wanted to be close to home.

At about that same time, I met Dr. John Lerma. John was an affable man, obviously a very competent physician, and fascinated with my books about close encounters. Although he remembered nothing of the kind in his own life, he could entertain the notion that it might be real. He was open to the idea of doing a surgery to remove the object in my ear.

Dr. Lerma did the surgery in a small suite of offices in San Antonio. A nurse was in attendance. Anne made a video record of the whole procedure, so what I am about to relate is verified by the existence of this tape. In addition, Dr. Lerma describes the same things in interviews.

This is what happened. First, he located the object in the upper part of the pina of my left ear. It was a hard swelling, diagnosable as a probable cyst. Since it sometimes became irritated when I slept on it, there was ample medical reason to remove it.

He determined that the object was firmly fixed in place, then drew a line around it. He prepared my ear with a swab of disinfectant and injected lidocaine anesthetic. The ear became numb, and the surgery proceeded. I was fully conscious the whole time.

I felt pressure as he worked, and a certain level of pain. I concealed my worry—even fear—with chitchat about baseball. He dissected down to the object, which he observed to be, in his own words, "a white disk."

When he touched this disk with the edge of his scalpel, a strange and unexpected event occurred. He observed that it ceased to be firmly fixed to the cartilage, as it had been before he made the incision. He reported that it moved away from the touch of his scalpel. When he touched it again, this time nicking it, there was a more decisive movement. The object relocated itself a full inch away from the incision.

Dr. Lerma became very silent. I tried to maintain a façade of levity by saying, "It's on the run." But I wasn't laughing. Inside, my heart was quaking. I can only describe what I was feeling as a kind of claustrophobia. I felt trapped, helpless, not a little desperate.

Dr. Lerma concluded that this unusual event made any further attempt to remove the object unwise, and he closed the incision. He had collected only a few fragments of the object. He placed these in a preservative solution and gave them to me. Most of the object remained in my ear.

I can remember how I felt as I got up to leave. "It's part of me, and it's somehow alive." These were my thoughts—my terror—at the time.

Anne and I went home; there she would sit close to me on the couch, quietly holding my hand as we watched television. Anne has been with me every second of all the years I have been having close encounters. She knows the terror, the anguish, the incredible, heart-crushing loneliness that we witnesses feel when we face the unknown, or the jeering, ignorant press, or scientists too deeply frightened to use what tools they possess on our behalf.

She knew what I was feeling, and she offered her strength and wisdom. I pulled myself together, and the next morning Bill Mallow and I were off on another scientific high adventure, putting the fragments Dr. Lerma had collected through their paces at the lab.

Bill reported two findings. One biologist said that the fragment appeared to have tiny tentacles on it. Another said, "It's just distressed collagen." Under the scanning electron microscope, the object was found to be filled with crystals, possibly calcium carbonate.

Beyond that, there could be little analysis, as there simply was not enough material in the small fragment Dr. Lerma had managed to collect.

I had to learn to live with my object. It still sits in my ear where it was, having returned to its original location about a week after the surgery. It still turns on and I still don't know what it does.

Dr. Leir was the man who inspired me to go ahead with the surgery. I'm not sorry that I did it. On the other hand, I'm also glad in some ways that the object is still there. I have become reconciled to it. Now, when it turns on during meditation or prayer, which it quite often does, I find that the sound doesn't intrude, but acts as a sort of companion to my concentration.

Maybe I am simply putting the best face on a bad situation because there's nothing else to be done about it. Or maybe the implants are not bad, maybe they represent an intrusion on our behalf by a more advanced species, and serve the welfare of mankind.

But what if, in our earlier example, we had gone into an animal testing lab instead of onto the African veldt? While the lion living out her implanted life on the veldt is contributing information that will help scientists preserve the species, the same cannot be said for the rabbit in the cosmetics testing facility. That animal is there for one reason only: to serve the needs of humankind, and not even for a very high purpose. Chemicals are implanted under his skin in order to see whether or not they irritate him—all so that we can use them in beauty applications without fear of harm.

So the rabbit lives and dies for our benefit only, as does the mouse implanted with cancer or the chimpanzee with brain electrodes.

The question has to be asked: Are we the beneficiaries of implanation, or its victims? In this book—and it is an extraordinary story, being told by a genuine pioneer—Dr. Leir does everything now possible to approach a meaningful answer to this question.

Somebody created these implants. Somebody wanted them there and had the ability to penetrate into the bedrooms of vulnerable people and put them into their bodies. Mine was placed under my skin without so much as the trace of a scar. Many of the others are the same: there is no sign of entry. In other cases, there are small scars—but carefully made ones, not the rough gouges that would be associated with accidents.

The conclusion cannot be denied, not by a rational and emotionally stable human being; Dr. Leir has made his case: Implants are real.

Are they something good or something bad? Should I rejoice in my good fortune or suffer my implant? Should I have it once and for all removed and damn the scars and damn the risk?

In this book, Roger Leir, for the first time, addresses these burning questions in a scientific way, and draws some unforgettable conclusions. If science could face the reality that somebody unknown—possibly even of human origin or somehow operating within the confines of government—is doing this to us, then some progress could be made. We might conceivably be able to make some intelligent decisions about implants. If science devoted enough of the right resources to the

questions surrounding implants, they could undoubt-
edly be answered.

I would like an answer, and if one comes, it will
happen because Roger Leir made it happen. Because he
has faced the reality of the implants—a reality that
makes lesser scientists sneer in open contempt and secret fear—we have a chance to take the next step. We
have a chance, truly, to understand.

THE BEGINNING
C H A P T E R   O N E

MY intention in writing this book is to provide a clear, concise account of how I applied my medical skills to a phenomenon that has not been looked at seriously by the scientific establishment. By doing this, I think I have come up with some of the most astonishing findings of any UFO research to date.

Fifty years after the modern UFO era began, mainstream science and medicine still regard the UFO phenomenon as foolish pseudoscience. If UFOs were taken seriously, there would be multimillion-dollar funding of a well-planned and coordinated global research effort. Instead, ufology gets along as best it can with minimal funding, raised largely by public membership in various UFO organizations and by small grants to individuals from daring philanthropic individuals.

The story of my own research efforts is one of co-incidence, curiosity, and struggle against many obstacles. This is the story of how a doctor and scientist got interested in the field of UFOs.

I was awarded the degree of Doctor of Podiatric Medicine in 1964. After graduation, I trained in surgery, and became podiatric director of residency training at Simi Valley Doctors Hospital and chief of the Diabetic Foot Clinic at Cedars of Lebanon Hospital in Hollywood. During this period I opened a private practice, where I still work today.

My interest in ufology dates back to July 1947. I can vividly recall my father walking into our kitchen and announcing to my mother that the United States Army Air Force had just captured a flying saucer. He was referring to the famous UFO crash at the air force base in Roswell, New Mexico. He showed her the newspaper headline and proceeded to explain his views on the subject of UFOs, discussing in depth his belief in extraterrestrial visitors. He also expressed his opinion that the government was keeping the phenomenon secret.

I have never forgotten his sincerity and the passion he had for this subject, so years later, when a friend asked me if I would be interested in attending a local MUFON (Mutual UFO Network) lecture, I told him I would, and accompanied him to the next meeting. The presentation was so interesting that I decided to attend future lectures, and I eventually became a member of MUFON.

Coincidentally, at this time my close friend and first cousin, Dr. Kenneth Ring, had just finished his book

The Omega Project. I was amazed to learn that he had
written about the link between UFOs and near death
experiences, or NDEs, because I had personally experienced an NDE that challenged my scientific assump-
tions and opened my mind to concepts such as
extraterrestrial visitations. Ken had discovered that both
abductees and those who returned to life, after a traumatic experience or during an operation, saw the same
beings and had some of the same experiences and after-
effects. Because of my new interest in UFOs, I found
this fascinating.

My NDE occurred on August 16, 1973. My recollection is vivid, because this incident was so traumatic
that it changed my life.

On a warm summer evening my close friend Jack,
my wife, and I arrived at Van Nuys airport about 5:30
P.M. We were going to fly to Bakersfield for dinner. I am
a licensed pilot, and had made arrangements with the
fixed base operator to have an aircraft ready and waiting. We parked, locked the car, and walked to where the
FBO had promised to leave the key and the logbook for
the aircraft. We made our way to the parked airplane
and proceeded with the preflight inspection. To my dismay, I found the fuel tanks almost empty. Never before
had I rented an aircraft with so little fuel on board. I
went to the nearest pay telephone, called the FBO at the
other end of the field, and arranged for the gas truck to
meet us. We climbed into the plane. I started the engine, called Ground Control for clearance, and taxied
toward the north end of the airport. In a few moments
we arrived there, met the waiting truck, and issued instructions for the type and amount of fuel needed. At

that point I made my way to the pilots' lounge and
checked the weather for our visual flight rules from Van
Nuys to Bakersfield. The weather was reported as clear,
with light winds and unlimited visibility.

The flight would take about an hour. I considered
this a short hop, one I had done many times before. My
passengers and I climbed back into the waiting aircraft,
secured our seat belts, and settled in for a pleasant,
routine dinner-hour flight. Jack sat to my right, in the
copilot's seat; my wife sat in the right rear seat. The
preflight check had been carried out. I started the engine and tuned in the Van Nuys ATIS (Air Terminal
Aviation Service) radio station. When I had the necessary information, I changed frequencies and called Van
Nuys Ground Control for clearance to taxi to the active
runway. We proceeded with our taxi roll to the run-up
area just short of the active.

All that was left to do was to finish the preflight
checklist and the final engine run-up. One by one I went
carefully over the list. All lights were on, all instruments checked and set. We were ready for takeoff.

Night was rapidly falling. I called the tower for
take-off clearance. Permission was given, so I taxied
into position on the runway and slowly advanced the
throttle until the engine was running at full take-off
speed. I carefully pulled back the stick, and instantly we
became airborne. After we started to climb, I commented to my passengers that it appeared we were going to be late for dinner.

The atmosphere aboard was relaxed, the departure
uneventful and routine. I headed the nose of our aircraft
directly toward the predetermined compass heading. I

set the navigational side of the radio to the halfway point of our destination. The needle slowly began to center and at a cruising altitude of ten thousand feet I leveled the nose of our little bird for a calm and smooth flight.

Everything was functioning normally. It was the beginning of a beautiful flight. We were starting to see lights below as the dusk melted into the blackness of night. The panorama appeared crystal clear, just as the weatherman had predicted. I asked Jack and my wife if they were enjoying the flight. They seemed relaxed and in awe of the beautiful scene below.

Time passed quickly, and soon the needle on the Omnigator started its slow swing from the indicted "to" to a "from" configuration. I reset the radio to a Bakersfield frequency, but nothing happened. This didn't disturb me, because there had been previous times when this had happened. I simply tuned Van Nuys back in and continued on my original heading toward Bakersfield. But then, much to my surprise and dismay, I realized we were experiencing complete radio failure. The only noise was a continuous popping sound over the speaker system.

It was now evident that we had lost all communication. I looked ahead and still could not see any light emanating from the Bakersfield area. This was strange, because at our altitude it should have been visible. "Perhaps there's ground fog in Bakersfield," I thought. If that was the case, I did not want to attempt a landing without radio communication with the ground.

To my right was the pitch blackness of the desert. There was no moon to guide us. I noticed that two beacons were visible, and in an instantaneous decision I changed course and headed toward those lights. I gave a short explanation to my companions and tried to reassure them. I simply told them that we were having a navigational problem and would be changing destinations.

Soon we were crossing the rugged Tehachapi Mountain Range in the cold night with absolutely no visual reference below, knowing there was desolate desert terrain ahead.

Suddenly the aircraft was gripped in violent turbulence. Everything in the cabin began to float or stick to the ceiling, including my charts and other navigational aids. I desperately fought the controls, trying to stabilize the aircraft. My eyes glued on the instruments, I concentrated deeply. All at once the gyrocompass started a slow roll and all attempts to stop the malfunction proved fruitless.

"My God," I thought, "what more can happen?" I wondered if the engine might be next. I decided to fly toward the beacon to my left, thinking it would be China Lake Naval Air Station. The light to my right would then have to be the small city of Inyokern.

To my left I could see a single row of runway lights in the blackness of the land below. I couldn't tell which side of the runway was represented by the lighting pattern. There were no taxi lights, no background lights, and no threshold lighting system. There was a rotating beacon, which appeared to be coming from the tower.

I took up a new heading directly toward the tower. Next, I executed the emergency signal, which consisted of turning my landing lights on and off in rapid succession and waving my wings. I peered out of my little left-hand window, looking for some response from the tower. It was only moments before I got my first surprise. It came in the form of a steady red light. This meant, Do Not Land!

I was shocked, and wondered what would happen next. Fortunately I kept my eyes glued on the tower and soon saw a steady green light shining at me.

At that point I informed my passengers of our true situation. I told them that we were attempting a landing at an alternate airport due to radio failure and other instrument problems. They had confidence in me as a pilot and responded to my directions without argument. I told them to make sure their seat belts were firmly fastened and to be prepared for an abrupt landing. They became quiet and asked no questions.

I was now cleared to land. I started my downwind approach, using the one row of lights below as a guide. Without a radio I was unable to set my altimeter, so I had to guess at an approach of one thousand feet above the ground. What I could not know or see was that the desert floor was closer than I realized.

The desert night air was hot, and I could feel the radiant heat from the ground below. Next I slowed the aircraft to 120 miles per hour and put down twenty degrees of flaps. I lowered the landing gear and waited for the indicator light to show that the gear was down and locked. My prop was set at full pitch and I maintained what I thought was an altitude of a thousand feet above the ground. I proceeded with a shallow base turn. Instantly my landing lights illuminated a desert floor of sand, cactus, and sage. Panicky thoughts flew through my mind; an extreme emergency situation had just begun. I knew that if the plane hit the ground with a nose-down position we would all perish. My next actions were purely automatic behavior.

My right hand shot toward the throttle control and with one swift motion I pushed the throttle to its maximum full-open, "fire-walled" position. I pulled back on the yoke and assumed straight and level flight. This was an effort to prevent a "stall" configuration. I knew that we were going to hit the sand, so I wanted the nose of the aircraft as high as possible. I kept the stick pulled back into my lap and watched in horror as I saw the left wing impact with the ground. We were instantly propelled forward, sliding in the hot desert sand, listening to the sound of tearing metal all around us. The nose of the aircraft came down with a tremendous jolt and I watched as the mighty three-bladed propeller took a bite out of the earth, which ripped the entire engine from the fuselage. We came to rest only five hundred feet from the distant edge of an isolated military runway.

I felt the jolt of the crash and then everything went black. Suddenly I felt myself rushing toward a tunnel of light. A feeling of absolute calm was upon me. I was surrounded by an aura of blissful peace. I was not far along on this journey when I heard a stern, commanding voice ordering me, "Go back!" There was the implication that I should have known better than to proceed and that I did not belong here.

Shortly after this directive, I regained consciousness and found myself rubbing my eyes, thinking I was blind. I rubbed away some of the blood which had filled my eye and realized that I could indeed see. Then I knew that I was staring out a blood-smeared window. I was still alive—and so were my wife and friend. We were taken to the China Lake Naval Air Station Dispensary, where I was diagnosed with contusions of the head.

Recently I discussed this NDE with a colleague who is familiar with near death experiences. He asked me how long I thought it took for all the events of the crash to occur. I pondered the question for a moment and answered, "Probably only a fraction of a second." He then asked me how I was able to recall the events that happened at the moment of impact so well—it was as if I were seeing everything in slow motion. It didn't take me long to realize that this was a physical impossibility. He told me that this was what is known as an "angel experience." As I reflected on it, I could not disagree.

Ever since that time, I have found myself in disagreement with colleagues who insist that such things as psychic communication, alternative healing, and UFOs can't be real. I have personally experienced the miraculous, and I know that we scientists have not discovered everything there is to know. That's why, when I was confronted with the evidence that objects were being implanted in the bodies of abductees, I was able to address the question directly. I saw no need for arguments between the scientific and UFO communities. If implants are being put into human beings, I decided, why can't we take them out? That way we can prove something real is going on.

THE MEETING
C H A P T E R   T W O

JUNE 1995 was filled with UFO projects. It seemed as though I was spending more time on ufology than on my profession. My phone was ringing off the hook. Most of the calls were from MUFON members asking me if I could do one thing or another. I tried to accommodate everyone who called, but it didn't take me long to learn I couldn't.

My practice was feeling the effect of managed care, with a decline in the number of private patients and a significant income loss. I thought about what I might do to supplement my earnings, and decided to write articles on podiatry for medical publications. I had no idea then what this would ultimately lead to.

It was during this period that the local Ventura–Santa Barbara MUFON group changed its monthly written publication to a new format. The editor asked me if I would be willing to become one of the contributors. This interested me, and I accepted the position.

I now felt an obligation to attend as many lectures as possible. I reported on these events, carefully presenting the facts as I saw them. With my interest growing, I decided to study for the position of field investigator. Within a few months I passed the examination and was officially working for the national organization.

In June, I received word that UFO Expo West was going to be held in Los Angeles. The literature was intriguing, so I made arrangements to attend. I would be attending as an enthusiastic ufologist as well as an investigative reporter. I could look, listen, enjoy, and be critical all at the same time. Alice Leavy, the Ventura MUFON section director, accompanied me on the trip. Alice is a sincere, dedicated individual who has had many personal UFO experiences.

We arrived at our hotel late in the afternoon, before the lectures began. This allowed us to get settled in our rooms and to take a look at the exhibits. There were many types of vendors present, about 90 percent of them involved in the field of ufology, selling videotapes, T-shirts, jewelry, dolls, and publications. As we walked through the display areas, we were greeted by many familiar faces. Some people seemed surprised that a member of the medical profession would spend time at a UFO convention.

The next day I got up early so I could attend as many lectures as possible. During my breakfast alone in the coffee shop, I planned out my day, using the conference program as a guide.

I took notes during the first lecture, to give myself a good start on the article I would write. Afterward, I ran into Alice. The next lecture was by a Texan named Derrel Sims. Alice asked if I knew who he was.

"Yes, I've heard about him, but have never had the opportunity to see him in person," I answered. "I believe he speaks on the subject of alien abductions. Would you like to attend his lecture?" She nodded, so we headed toward the lecture hall.

The room was quite large, and along the walls were tables with various displays of pictures, books, and other UFO materials. I noticed a long table displaying several photographs and what appeared to be medical and scientific data. "Hey, Alice! Come over here and take a look at this," I said. We approached the table together. The first item we saw was a display of foot X rays depicting some sort of foreign object in the big toe area. Standing by the table was a tall, good-looking man engaged in conversation with several other people.

He turned and looked at Alice. "Hi, my name is Derrel Sims," he said, stepping forward to shake hands with her.

She replied, "Hi, it's very nice to meet you. I'm Alice Leavy, the section director for the Ventura–Santa Barbara chapter of MUFON. I'd like you to meet Dr. Roger Leir. He's our medical consultant."

Derrel stuck out his hand and gave me a hearty handshake. "I'm very happy to meet you," he said graciously.

I gestured toward the X rays. "Do you mind if I pick these up and have a look at them?" I asked.

He reached over and handed me the X ray films in one swift motion. "Take as long a look as you want, Doc."

As Derrel and Alice resumed their conversation, I held the films up to the light and scrutinized the foreign objects. My initial impression was that they looked very similar to stainless-steel sutures commonly used in foot surgery. I thought, "This is a total waste of time. I'm not giving any more attention to some guy who displays X ray films of foot surgery and makes a big deal out of the whole thing." I gently laid the films back down on the table and started to walk away.

Then I heard Alice's voice calling me to return, so I moved back to the table. She grabbed my arm and said, "I think you should listen to what Derrel has to say. It's really very interesting."

I gave her a nudge and as she moved toward me, I whispered in her ear. "Alice, come on, let's not waste any more time with this guy. It's a bunch of non-sense—just someone who's had foot surgery and is claiming some absurd story about alien implantation."

She looked at me as if I had just come down with the Plague. "Roger, you aren't being fair. You should give Derrel a chance to tell you what this is all about."

"Okay, I'll listen to what he has to say. But re-member, I'm only doing this because you asked me to," I said grudgingly.

Derrel was now surrounded by a large crowd of people who were listening intently to him. We edged our way closer and at the appropriate opportunity Alice

jumped into the conversation. "Derrel, please tell Dr.
Leir what you were explaining to me about this case
with the foot X ray."

I listened closely as Derrel began to relate the story
of an abduction history and the events that led up to an
X ray being taken. I asked him if the person involved
had had foot surgery, and he said none had ever been
performed. I then asked if he could present proof of his
statements, and to my surprise he led me to a gigantic
pile of folders. Leafing through several stacks, he came
up with a thick medical file on which the name of the
patient was covered over. "These are her medical
records. Would you like to go through them?"

"Sure, I'd love to take a look," I answered. He
gave me the file. "Would you have any objection if I
took this data to my room, where I can go over it in
peace and quiet?"

He responded by handing me another stack of documents, saying these contained additional information
and reports from the University of Houston. I thanked
him, put the package of documents under my arm, and,
with Alice following, walked briskly away. We pushed
our way through a small crowd of people gathered at
the doorway and proceeded into the hall. "Well, that
was quite an experience," I commented. Alice agreed
and suggested we take a break for lunch.

When lunch was over, we sat looking once again at
the lecture and workshop schedule.

"I see that Derrel is going to be giving a lecture at
three this afternoon. I'd like to attend. Wouldn't you?"
Alice asked.

"I guess we could do that," I responded.

When we reached the lecture room where Derrel was speaking, it was crowded with people. We managed to find two seats near the front of the room. Derrel entered from stage left and took his place at the lectern. First, he described his background and mission in life. I pulled a pen from my pocket and began to take notes. Derrel told us that he investigated abductions for the Houston UFO Network in Texas. He himself was an abductee who had spent nearly thirty years gathering hard physical evidence on the abduction phenomenon, including a variety of alleged implants and an alleged artifact from the UFO which crashed near Roswell, New Mexico, in 1947. He had extensive training in clinical hypnotherapy. For years, Derrel provided hypnotherapy at no charge to people involved in alleged abductions. Derrel went on to talk about the abduction cases he had investigated over nearly three decades.

I had dinner with Alice and some of the other attendees, and by 10 P.M. I was eager to return to my room. At last I was alone. I wanted a chance to look at the large amount of material I had gotten from Derrel. The more I read, the more intrigued I became. There was no medical evidence to suggest this individual had ever had foot surgery. The records also contained a report from the physician who took the X rays. He stated that when the patient had been questioned regarding the objects in her foot, she had denied having ever had any medical or surgical intervention whatsoever.

In addition to the medical records contained in the file, there were other documents relevant to this and other cases which Derrel had been investigating. One document was a report by a professor in the University of Houston Physics Department about an analysis performed on an object submitted by Derrel. This object was obtained from one of his abductee clients, who reported that it came from her eye. The in-depth report was quite impressive. Attached was a photograph of the object taken with an electron microscope. The analysis showed a number of unusual properties. The object had a hard, ceramic outer cover and a soft, velvety interior. There was speculation that it might be the housing for a biological camera installed in the client's eye during an abduction episode.

Finally I turned out the light and attempted to drift off into a peaceful sleep. But I was too restless; my mind was filled with thoughts of the day's events. I kept thinking about abductions, implantations, and alien entities.

The next morning I met up with Alice in the coffee shop and we mapped out our strategy for the rest of the day, deciding to attend some of the lectures together. We planned to meet at up at Derrel's workshop that afternoon.

The day passed quickly; both my mind and my yellow notepad filled up with information. I rushed down the hall toward the room where Derrel's workshop was scheduled. I could tell by the crowd gathering at the door that this was going to be a packed room. A young man was standing at the doorway taking tickets for the event. He asked me for my ticket, and I told him that I didn't have one. Then Derrel appeared just as Alice arrived.

"It's okay, they're my guests," he said, escorting us to seats at the front of the room. I told him I'd gone over the abductee's medical records and found them quite interesting. I asked why she didn't just have the object removed and analyzed. He explained that this person had no medical insurance and couldn't afford to have the surgery performed.

I thought deeply for a moment. "Derrel, would the lady be willing to come to California to have the surgery performed?"

He looked at me and said, "I don't see any reason why she wouldn't."

"Well, I'll tell you what," I responded. "If she would be willing to come here, I'll perform the surgery without charge."

"Do you really mean that?"

"I certainly do," I quickly replied.

Derrel shook my hand vigorously.

During the course of his presentation he mentioned the case that he had discussed with me, explaining that I had just offered to perform the surgery without charge. He said the woman couldn't afford the airfare from Houston to California and he asked anyone in the audience who might be willing to help pay for it to see him afterward. Then he finished his presentation.

As the crowd was leaving, I heard a voice over the loudspeaker system say, "Dr. Leir, please come to the podium. Dr. Leir, please come to the podium."

Holding Alice by the arm, I forced my way through the crowd. At the podium, Derrel was talking to someone. Unsure of why I had been paged, I went up to them. Derrel looked at me, beamed a huge smile, and said, "Dr. Leir, I'd like to introduce you to a gentleman who has just agreed to foot the bill for the airfare."

There were many times in the following days when I thought about the situation, worrying about what I might be getting myself into. One of the requests I had made was for Derrel to send me the X rays of the patient, so I could take them to a qualified radiologist prior to performing any surgery. I also explained to him that the surgical candidate would need to have laboratory examinations done prior to coming to California and that I would supply him with the orders for the type of lab work needed. In addition, she would be required to start on antibiotics two days prior to the surgery. Derrel agreed to these conditions. I was ready to embark on my first trip toward the unknown.

THE PLAN
C H A P T E R   T H R E E

ONLY a few days after I returned from the UFO conference, the phone rang one morning at 7:45 A.M. I am a night person and my office hours are tailored to my lifestyle, so I was jolted awake. I picked up the receiver. "Hello," I said hoarsely.

"Hi, Doc! This is Derrel, in Texas. How are you doing this fine morning?" My mind not yet fully conscious, I thought, "Who in the hell is Derrel?" At that point my memory began to function. I realized that the two-hour time difference between Texas and California might become a problem for me because of my new-found colleague in Houston.

I responded, "I am just fine. How is every little thing in the great state of Texas?" Derrel seemed to pick up on my mood.

"I hope I didn't get you up."

"Oh, no," I lied politely. "I was just heading for the shower." I told him I would get back to him later that day.

It was a typical day at the office. But by the afternoon, I was actually ten minutes ahead of schedule, so I picked up the phone and called Derrel. He got right down to business. "Doc, have you given any more thought to the surgery?"

I told him that I could speak more intelligently about the case when I had some information in front of me, and asked him to supply me with particular data about the proposed patient. He seemed satisfied with this, and we ended the conversation.

It dawned on me that I had pledged my services as a surgeon without considering an important factor: What about the criticism I might receive from my colleagues and peers? With a wife and a little girl to raise, I couldn't afford to give up my practice at this stage in life. Maybe, it occurred to me, I should have thought of that before I opened my big mouth. I came close to calling Derrel and canceling the entire plan.

At the end of the day, I called Alice. "I'm concerned that I might have taken on too much with the promise I made Derrel," I confided. "You know, I might get into serious trouble professionally by proceeding with this."

I explained that performing the surgery was one thing, but that the required documentation, if published, could damage my credibility in medical circles. She felt that I was overreacting and cited other professionals in MUFON who'd had no problems due to their

UFO-related activities. I tried to cast off my worries so I could get a good night's sleep. But that night I had some troubling dreams.

The next morning, in the shower, I noticed something unusual on the palm of my right hand. Rinsing off the soap, I saw what appeared to be a line connected by three symmetrical dots, extending upward across my palm to the third finger. I thought I must have marked myself with a ballpoint pen. Scrubbing forcefully with a washcloth and soap, I tried to remove the mark, but it would not come off. I figured that whatever had made the mark was insoluble in plain water and soap, so I planned to remove it with alcohol when I arrived at my office.

When I got to the office, I quickly donned my white smock and raced into the bathroom. I took a bottle of alcohol from the medicine cabinet and scrubbed the strange mark with a saturated cloth. To my amazement, the mark remained. I reached for a bottle of nail polish remover and scrubbed again—nothing happened. I asked my office manager, Janet, to bring me a large Band-Aid so I could cover up the mark. She returned shortly with a one-inch-wide bandage. I peeled the paper free and placed it over the odd markings on my hand.

"Where in the world did you get that weird mark?" she asked.

"To tell you the truth, I don't know. I just woke up with it this morning," I grumbled.

The rest of my workday proceeded without any more unusual happenings. But my decision to operate on Derrel's client gnawed away at the back of my mind.

I had to let him know something soon. I was about to
leave when Alice called to ask if I was going to be on
time for the meeting. I had completely forgotten about
that night's MUFON meeting. I told her I would get
there as soon as possible and to start the meeting without me.

When I arrived, the room was filled with people
and Alice was at the podium giving her welcoming
speech. It was usually my role to introduce the evening's speaker. I waited for my cue from Alice and
hastened to the podium. After I addressed the audience,
the speaker came up and I retired to the back of the
room.

Alice told me that one of our board members had
brought in some strange material which had appeared
overnight in the form of a ring on her front lawn. It was
later determined to be slime mold, an ordinary earthly
substance, but at that time it remained unidentified. Alice showed me the plastic bag and asked me to take it
out to my car and test it for fluorescence with the black
ultraviolet light I had there. I took the bag and headed
for my car, with Alice and a few others at my heels.

I opened the trunk of my car and took out the black
light. One of our group held the bag open as I shined
the light so all the material could be exposed. There
was no obvious fluorescence. Abruptly, Alice shouted,
"What is that all over your hand?"

I shined the light on my palm. The bandage I had
applied earlier was falling off. The area of the mark was
fluorescing a brilliant green color. I stood there in
shock. In his lecture, Derrel had described a brilliant

green fluorescence appearing on the bodies of approximately 5 percent of alleged abductees. Suddenly I remembered my uneasy sleep of the night before.

I arose the next morning with a nagging question on my mind: Had I become a victim of the very phenomenon I had recently become involved with? I immediately knew what I had to do.

The phone rang three times before Derrel's voice answered. I said, "Derrel, this is Roger. I've been considering the surgery, and want to start making preparations as soon as possible. I will fax you the information I require on the patient." I gave Derrel the details of what had to be done and some general instructions pertaining to the requirements.

Saturday morning found me sitting quietly in my office preparing to make the necessary arrangements for surgery on Derrel's patient. Almost as if he were reading my mind, Derrel called on my private line at that instant.

He asked me some questions in regard to the lab tests I wanted. I supplied him with information and then he sheepishly asked, "Doc, do you think that you could arrange to do another surgery in addition to the foot case?" He explained that a man who was an alleged abductee had some sort of a metallic object on the back of his left hand. I asked him if he had seen the X rays and he replied that he had. In addition, he had known this person for quite some time and was convinced of the sincerity of his assertions.

He also advised me that neither surgical candidate had ever undergone hypnotic regression regarding their abduction memories. I told Derrel I would attempt to arrange for another surgeon to do the hand case, since podiatrists are limited to foot surgery. I had to consider several possibilities: an orthopedic surgeon, a plastic surgeon, a specialist in family practice, a dermatologist, or a general surgeon. After careful thought, I decided to go with a general surgeon. My primary reason for choosing this specialty was the need to have someone who would be prepared to handle anything unusual.

I finally chose a friend I had known for many years. I shall refer to him as Dr. A. Not only was he a close friend, but his qualifications were impeccable. He was retired and would do the surgery without charge.

Next, I considered the need for a psychologist, in case these patients were disturbed by their abduction experiences and needed counseling before or after the surgery. I turned to the list of members in our local MUFON group and came up with Christie, who was extremely gentle with her patients and always willing to learn. I knew she would be thrilled with the opportunity to participate in the project and would not ask for payment.

It was also important to have an attorney on our team. Since the entire procedure would be documented for research, all the paperwork, such as consents and releases, needed to be handled in the correct manner. I had known many attorneys over time, but most of them would not perform services without charge.

One likely individual did come to mind, however. Jerry was more than just an attorney; I had known his family for many years and had worked with him on several legal matters. I knew I could approach him about the project. I also knew he might laugh at the whole idea but that if he thought it could possibly be fun, he would throw himself into the project whole-heartedly.

Derrel would perform during the surgery as hypno-anesthesiologist. I had decided to perform one case with the use of hypnosis and one without. This would add another medical research factor to the picture. We would then be able to do a comparison study in reference to healing, reaction to medication, and the requirement for postoperative analgesics. This would be independent from the alien abduction research component.

After the principals were selected, I turned my attention to the remaining portion of the team. I realized that my surgical nurse should be someone who was not only familiar with my surgical techniques but also a person I could trust completely. After careful consideration, there was only one person who would fill the bill: my daughter-in-law, Denise. Before giving birth to my granddaughter, Denise was a trusted employee in my office, and had given up that position to become a full-time mother.

My next concern was how to record the event. This aspect was important because it was, above all, a research project. I decided to use both still photography and video. In addition, written documentation would be necessary.

I needed a very special person to operate the video camera. I did not want an individual who would become shocked or revolted by the sight of blood and cause a disaster by fainting directly into the surgical area.

I pored over a list of candidates from among the

MUFON members. All at once, the name Mike Evans popped up. He would be perfect for the job. Mike was a registered nurse employed by the Veterans Administration, and was responsible for video-recording all of our local MUFON meetings.

I needed someone to make a written record as well. I picked Jack and Ruth Carlson. Ruth was past president of the local writers club. Jack had written many articles for the Ventura MUFON newsletter.

I asked Bert Clemens, who is a general contractor with skill in electronics, to make sure all the equipment functioned properly. I decided to transmit the event to an adjoining room where guests could witness the surgery independently from those who were present in the operating room itself. In order to accomplish this, cables had to be connected from the camera in the operating theater to the adjoining room with the television receiver. I knew this would be an easy task for Bert.

Most important of all, I asked Alice Leavy and my office manager, Janet, to coordinate the entire event. At that point, I was satisfied with my personnel choices.

Next my thoughts were drawn to the selection of a suitable surgical site. The apparent choices were a hospital, an outpatient surgical center, or one of my offices. I had to consider that a hospital would provide a better facility with more equipment, but the number of people involved in the project was a major drawback. Also, the television equipment would be cumbersome and I might need special permission in order to use it. Using an outpatient surgical center presented similar problems. The only logical alternative was one of my three offices. I decided my Camarillo office would work best.

I shared this facility with two dentists. If I performed the surgeries on a day when they were not in the office, it would provide us with the ideal location. A single phone call would settle the matter, so I dialed our number and reached the answering service. As I waited for one of the dentists to come on the line, I wondered what to tell him. I didn't know just how far I could go with him, because we did not know each other on a social basis. I thought it best to simply ask about the surgery and leave it at that.

"Hello, this is Dr. King. Can I help you?"

"Larry, I was wondering if there would be any problem with me performing a surgery in the office one Saturday. I know that you and Dick usually don't come in on weekends to see patients."

"I really don't see any problem with that. When do you think this would be?"

I realized I was going to have to nail down a date, right then and there. July was sailing past rapidly. I flipped through the calendar and picked August 19. My mind raced. Could I possibly get everything ready by that time?

At that point, I realized that I had committed myself to the project.

That evening, after three hours on the phone calling all the people I had picked to volunteer, I was all fired up and ready to push onward. I was still at the office and I had completed all the calls but one. Absolutely nobody had turned me down.

The only person I had left to call was Dr. A. I'd left him for last because I knew his schedule was a little erratic. When I reached him, he greeted me warmly.

"What are you up to this time of the night?" he asked. "I figured you'd be home with your wife and kid."

"When you hear what I have to say, I think you'll understand why I'm still at the office," I responded.

"Okay, let's hear what's got you so excited this time."

"I'm going to tell you something, and I want you to promise you won't laugh."

Dr. A. assured me that whatever I had to say, he would take it seriously. I explained to him that I had gotten interested in the field of ufology. At that time, even some of my closest friends weren't aware of this. I told him how I'd met Derrel and how this had led up to the impending surgeries.

The tone of Dr. A.'s voice became quite serious as he asked, "How do you know you can trust this guy? And how do you know these patients will even show up?"

"Actually, I don't really know Derrel at all, but I've decided to take a chance on him. Also, my friend Alice, who is the head of MUFON out here, seems to have great confidence in him," I answered in one breath.

"Well, if you think this is actually going to happen, I'll help you as much as I can. What date is this event scheduled for?"

"I have it marked on my calendar for the nineteenth of August."

Dr. A. said that he would set that day aside and wanted to get together with me several times before the event to discuss the matter in greater detail. He asked to review the medical data as well as to preview the X rays.

During the next few days, I had time to consider some of the unknowns regarding the upcoming surgeries. I certainly didn't have to consult the books regarding procedures for removing foreign bodies; I had performed this type of surgery hundreds of times before and had always had great success. During my years of practice I had removed many different objects from the human foot, such as shards of glass, wooden splinters, pieces of metal, needles, plastic, and even hair. My concern was that I knew nothing about so-called alien implants. I decided that my collection of UFO literature must contain some information regarding the subject.

I pored through the literature, but couldn't find much information.

Budd Hopkins, in *Intruders,* recounted that a physician had removed a small BB-like object from the nasal passage of a woman and then simply thrown it away. I learned of another woman, named Cheryl Fernandez, who supposedly had an alien implant surgically removed from her leg. In addition, there was a Canadian physician who reportedly removed an object from an individual with an alien abduction history. I could find no published analysis of any of these objects.

One of the most notable cases involved an alleged abductee from New York State who had an implant in his penis. Over time, the object emerged on its own and the abductee was able to remove it himself. He then offered it to Dr. David Pritchard of the Physics Department at M.I.T. Dr. Pritchard initially examined the object by electron scanning microscope, but did not find any unusual compounds or persuasive evidence that it was of extraterrestrial origin. These tests were financed by Robert Bigelow of Las Vegas, who was apparently dissatisfied with the results and persuaded Dr. Pritchard to do some further testing. The final conclusion was that the object was interesting, but unknowable.

Faced with tales of frustration and failure by others in their quest to obtain hard evidence of alien implants, I knew I had to take every precaution in order to prevent similar problems.

I also read about how the few objects that had been successfully removed had had strange things happen to them, which was one reason for their lack of analysis. There were tales of surgically removed objects turning to powder, becoming liquid and vaporizing.

I decided that my objective was not only to extract these objects, but to preserve them so that they would not rapidly degrade. I gave this matter long, hard thought and came up with a plan. The safest medium in which to place one of these specimens would be a biological fluid belonging to the patient from whom the object was extracted. The simple answer dawned on me.

All I had to do was to withdraw blood from each one of the patients, spin it down in the centrifuge, and separate the serum. The serum would be mixed with a preservative which, without the presence of blood cells, would not coagulate. This would become the ideal medium in which to store the implants after they were removed.

   The next few weeks were filled with meetings and
phone calls from all the parties involved. As the day of
the surgery approached, all the pieces started to fall in
place. I was confident that I was ready for my journey
into the unknown.

THE ABDUCTIONS
C H A P T E R   F O U R

ONE of the criteria we established for removal of alleged alien implants was an abduction history of the proposed surgical candidates. Derrel took responsibility for supplying a detailed abduction history for Patricia and Peter, the surgical candidates.

Critics and debunkers of the alien abduction phenomenon point disapprovingly to the methods of investigators. Their number one criticism focuses on the use of hypnosis. In some instances, their criticism has merit. This is because no hypnosis methodology has ever been universally adapted by researchers. Although there are many well-qualified individuals performing hypnosis, there has never been a professional standard developed for the best method of memory retrieval. In a world full of complexities such as false memory syndromes, childhood abuse memories, and past life regressions, we must adopt research criteria that are beyond reproach. Therefore, after careful consideration, Derrel and I decided to forgo the use of regressive hypnosis altogether. The abduction histories of our surgical candidates that are presented here are based on as many details as they could recall without the use of hypnosis.

In 1969, Patricia and her family lived in a small rural area of the state of Texas. She was twenty-three years old and married to John, a caring man with a very strong personality. They were the parents of two boys, Michael, six years old, and Billy, five. John had been hard at work for months without a break and Patricia felt they were due for time away from the daily routine. She also didn't have much more time left for a vacation, since she was eight months pregnant.

One evening in October, while Patricia bustled about the kitchen preparing dinner, John came in smiling. "Honey, how would you like to take a little vacation?" he asked. Patricia was enthusiastic.

Several days went by before the subject came up again. John began by telling Patricia that he had discussed the idea with a couple of his buddies and they'd suggested an excellent place to go fishing. Patricia considered the idea for a few moments and said, "You know, honey, that really sounds like a great idea. I'm sure the boys would just love it. We could get all the camping gear out and really have a great time."

He said he'd tell the boys about the trip. A few moments later, two bright-eyed boys burst into the kitchen, shouting, "Mom! Are we really going on a camping trip?" Patricia assured them they were.

The next evening, plans were finalized. It was decided that if they took the boys out of school for only two days, they wouldn't miss too much work. This would get them on the road the following Thursday.

On the morning of their trip, John put the car into gear and stepped on the gas, and the automobile lurched onto the highway. Their adventure had begun.

After what seemed like hours, Patricia looked at her watch. The trip was taking longer than expected; it was now almost 4:30 P.M. and they had not arrived at their destination. Some of the roads were not clearly marked and it had become increasingly evident that the map they were using was not up to date. "John, how much longer until we get there?" she asked.

John took a quick glance at the map and said, "We should just about be there."

Billy had his face pressed against the car window and was drawing figures in the steam created from his hot breath. Michael seemed oblivious to his surroundings as he flipped through the pages of an old comic book.

Patricia peered intently into the distance and noticed a structure ahead. "Is that the old iron bridge your friend told you about?" she asked her husband.

John stared ahead, then slowed down and said, "It sure is, honey."

This was the spot John's friend had talked about. It was an old wrought-iron bridge that crossed a beautiful stream, and had a rich Civil War history. The Confederate Army had built it as part of its supply route.

John slowed down as he approached the old bridge. The automobile tires made a creaking and thumping sound as they began to cross the wood planking. Patricia rolled down the window and stuck her head outside. "Do you think this bridge is safe?" she asked.

"My friend told me it was built to last forever," John replied.

Soon they were on the other side. There was a clearing to the left and a small trail that led down to the river. Billy and Michael were jumping up and down on the backseat.

"Hold on, guys, we're almost there," John said. He stopped the car at a small section of the clearing just after they passed the far end of the bridge. It was an ideal campsite, close to the river and below the level of the bridge. The family scrambled from the car.

It took about an hour to prepare the camp. Father and sons went in search of firewood. John had a good deal of experience in making outdoor fires and knew what kind of wood it took.

Darkness slowly began to consume the campsite. A few stars were clearly visible in the sky overhead. The boys went back to the car and brought out two small flashlights. They found a comfortable log to sit on and began flashing their lights at the sky.

"Hey, Mom, come on over here! I want to show you something," Michael yelled. Patricia walked over to where the boys were sitting, with John right behind her.

"Hey, look at this! It's really neat." Michael pointed his flashlight skyward toward a brighter than usual star, and made two short flashes. Patricia and John watched in amazement as the two flashes were returned!

"Patricia, did you see that?" John shouted.

"I sure did!"

John ran back to the tent and got two larger, more powerful flashlights. He gave one to Patricia and aimed his at the bright star, sending up two short flashes and then a long one. Within a few seconds, the same sequence of flashes was returned.

"Patricia! For God's sake, did you see that?" he asked excitedly.

She didn't answer, but instead held up her light and flashed a different series. These were also returned. Soon the single bright star was joined by others. The whole family was having a great time, flashing away and getting replies.

Later, the only sounds besides the crackle of the fire were night sounds of the open country. Both Billy and Michael were tucked securely into their sleeping bags. Patricia cleaned up the dinner dishes as John found a comfortable spot on a log and drank a cup of hot coffee. Patricia came over and sat beside him. He looked deep into her eyes and asked seriously, "What do you think was going on before with that weird light show?"

Patricia said, "I bet they were a bunch of helicopters from the air base. Probably some hotshot military air jocks out to have some fun. Well, what the heck! We had as much fun as the kids did!"

The following morning was filled with excitement as the lines hit the stream and the boys started to catch fish. The weather cooperated, the temperature staying in the mid-seventies. By noon, the family had caught its limit.

That evening, after a fish dinner, Patricia escorted the boys to the tent and tucked them in for the night. John puffed on his pipe and talked about his plans for the next day: to get up early and hike downstream to a spot where his friend said some of the biggest fish ever caught in the region had been taken. For that reason they decided to retire early. John checked the fire one last time. It was well stocked with slow-burning wood and would probably still be going when they arose in the morning.

Suddenly, at about 1 A.M., Patricia was awakened by her terror-stricken husband, who loomed over her like a huge giant, silhouetted in the reflection of the firelight. He pointed a finger at her and said, "Get up right now and throw the kids in the car!" His voice was gruff and demanding. Patricia's heart pounded and her mind raced. She wondered what could possibly be so wrong to cause her husband to act like this. She quickly got out of her sleeping bag and put on her shoes.

John was now outside the tent and she couldn't see what he was doing. A few moments later he stuck his head in the tent opening and bellowed, "Hurry up! Come on!" Patricia grabbed Billy and dragged him forcefully out of the sleeping bag, the sudden movement waking him up. He started to cry, so Patricia gently soothed him in a quiet mother's voice. "It's all right, dear, we're just going for a little ride." She rushed out of the tent and placed him on the rear seat of the car. "You go back to sleep now," she said softly. John appeared, carrying Michael, who was still sound asleep.

"John, please tell me what is going on," she pleaded.

His answer was a curt "Never mind now! Just get in the car!" She did as she was told. John immediately followed. He started the motor and put the car in gear, and they quickly began to drive down the desolate, dark road, leaving the campsite with their belongs still in it.

As they raced along the deserted road, Patricia was becoming hysterical. She pleaded with her husband to tell her what was happening.

He turned toward her with a look of terror on his face. "Something was under the bridge!"

"What are you talking about? Deer? People? Other animals?"

Her husband merely shook his head and said loudly, "No!"

John was concentrating on the road ahead. They were heading back the way they had come, toward the old iron bridge.

He suddenly turned to Patricia and said, "Look behind us! Do you see that bright light following us?"

She turned and peered out the rear window. "Yes, I do see a light."

John said, "Don't be frightened—I'll protect you. Maybe it's just a truck. Try and see what it is."

Patricia strained her eyes to see what was following them. "Yes, yes, I think it is a truck," she stammered. The light was getting brighter and brighter, beginning to match the speed of the car. She was gripped with cold terror as they approached the old iron bridge, a dimly lit silhouette in the darkness.

It was becoming apparent that the brightly lit object

behind them was not a truck, and in fact was of such a gigantic size that it would never fit on the tiny bridge. They were now almost across the bridge, just a short distance from the other side.

Patricia's attention was suddenly drawn to her husband. John looked like a zombie. "John! John!" Patricia yelled, poking her husband on the arm. "Stop the car right now," she commanded. John quickly stopped. There they sat, dumbfounded.

"What just happened?" asked Patricia after a moment.

"Honest to God, honey, I don't know," John replied. Together, they stared through the windshield in amazement. The car was now pointed in the opposite direction from the way that they had just been heading. The road before them led back toward the campsite, not away from it!

John stepped on the gas and the car once again clattered across the old, tired bridge, which groaned and creaked under the weight. As they approached the campsite, Patricia was aware that something was wrong. The first thing she noticed was that the campfire was out except for the red glow of the ashes. She thought, "This is impossible—John made that fire to last all night. We've only been gone a few minutes." It was as if hours had vanished. She looked at John, trying to read his thoughts. He was extremely agitated and afraid. Patricia knew her husband was a man who was afraid of nothing and no one. Just what had happened? The question began to eat away at her.

They stared at each other. Sharing the same thought, they both blurted out, "Let's go home!"

They packed up quickly. The boys appeared to still be sleeping comfortably on the backseat of the car. During the trip home, John drove too fast for the road conditions. Patricia was concerned about this, but kept silent. She turned and shifted her gaze to the road behind them, and noticed a greenish light following them. She was afraid to say anything to John, but from the look on his face she realized he must have seen it in the rearview mirror. The object continued to follow them for several more miles. Patricia looked over at her husband and noticed a tear running down his cheek.

He suddenly said in a loud voice, "Please, don't hurt us!"

The children were beginning to stir and seemed agitated and restless. Patricia gently pushed them back down on the seat. John continued to panic. He threatened to stop the car, allowing their pursuers to catch up. Patricia pleaded with him not to do it.

During the long trip home, fear was with them constantly. Time dragged as the greenish light kept pace behind them. Finally, they turned onto the main highway and John pushed the car to its maximum speed. Eventually, the light disappeared. It was almost 7 A.M. Soon they were looking at the most welcome sight they had ever seen—their home.

This ended the couple's strange camping adventure, but it was not the last of Patricia's experiences.

A few days following their return, Patricia was due for a checkup by her obstetrician. She was examined in the usual manner. She was told that the course of her pregnancy was proceeding normally and that she was going to have a healthy baby. And indeed, her daughter, Sonya, was born on schedule.

Patricia has always been concerned about the health and welfare of her daughter because of her experience on the camping trip. She says her daughter has grown into a marvelous human being. She is extremely intuitive and has an uncanny sort of wisdom. She has a high IQ and has become a skilled writer. Sonya expresses many premonitions about the future and is much different from Patricia's other two children.

When I questioned her about how the camping trip might have influenced her unborn child, she simply said that she did not know. The area of their encounter was later flooded by the Army Corps of Engineers, and there is now a lake at that spot.

It was a cold, damp night about a year later, in 1970, and Patricia and John decided to turn in early. They had many chores to do the next day and wanted to get some extra rest.

The small lamp on the nightstand produced a dim and relaxing glow. Patricia decided to read for a little while before going to sleep. She had only read a few pages when she began to feel drowsy and reached over to turn off the light. The room was totally dark except for the glow of the alarm clock, which indicated it was 12:05 A.M. She drifted off into a light sleep.

The clock read 3 A.M. when Patricia became aware of a strange greenish light shining on the bed. She turned her head and looked around. The light came from outside the bedroom window and filled every part of the room. Her free will seemed to disappear, and her body would no longer respond to her commands. She thought, "What in heaven's name is happening to me?" She tried to turn and wake up her sleeping husband but found that she could not move her body. "John, wake up! Something strange is happening and I'm scared," she shouted. John didn't stir.

After some time had passed, Patricia gained control over her body again. She turned her head and saw that somehow she was in a gigantic room. The surroundings appeared to be made of glistening chrome. She was sitting on some sort of table and all around her were strange beings; to this day, she cannot remember what they looked like. She gazed about the room. The only part of her body she couldn't move were her hands, which were being held by the beings standing beside her. Her eyes fixed on three huge transparent cylinders, apparently composed of clear glass, which she estimated to be about eighteen feet tall. Contained in each of these cylinders was a basketball-sized object similar to a ball bearing, which she heard making strange noises, and each tube had what appeared to be a huge fan belt that was twisted in on itself to form a continuously moving Mobius strip. The objects looked as if they were in perpetual motion. A sound like the roar of wind seemed to be coming from inside the cylinders.

Patricia noticed that she was on what appeared to be an upper landing or balcony, about thirty feet above the cylinders. She thought, "If I were to fall, I'd be killed instantly." She stared once again at the area below, taking notice of machines that looked like large gears, along with other equipment.

The next thing Patricia recalls is being back in her

bedroom, lying next to her husband. She called out once again, "John, please wake up!"

John sat up straight, a horrified expression on his face. He turned to Patricia and said, "What's going on here, and where is that green light coming from?" Before Patricia had a chance to respond, the room was filled with a humming sound which seemed to come from outside and above their house. The greenish light began to fade, and then the room was dark again, with only the faint glow of the alarm clock remaining.

Patricia never recalled any objects being implanted in her toe. She discovered them when she stepped on a sliver and went to the doctor to have it removed. He X-rayed her foot and noticed what he thought were two pins in her left toe. He asked her when she'd had surgery, and didn't believe her when she denied that she had ever had an operation on her foot. There was no scarring, and no indication that it had ever been cut open or that objects had been introduced into it. When she later heard about implants, she remembered her dream and suspected that this was how she had gotten the strange objects in her toe.

Peter's story is similar to Patricia's. In 1954, he lived with his family in the small, rural town of Dubberly, Louisiana, a hundred miles from Shreveport. He was six years old, and lived the simple life of a farm boy. Peter went to a rural school and was responsible for certain chores on the farm.

After a typical day at school, Peter arrived home one afternoon at about three. The family ate dinner, and once his chores were done, Peter prepared for bed.

Since it was customary for his family to get up early in the morning, they usually retired early, at about 9 P.M.

Later, Peter suddenly awoke from his sound sleep, immediately alert. He seemed to lose control of his body, as if it wanted to do things on its own without his telling it to. He found himself getting out of bed and walking toward the door of the bedroom, then continuing down the short hall until he reached the back door of their little house. He reached for the doorknob, opened the door, and peered into the blackness of the night beyond. He stepped out onto the porch. The old wooden boards creaked beneath him.

The young boy didn't notice the cold outside temperature, nor did he feel his bare feet touching the cold ground. Guided by some inexplicable force, he proceeded around the back of the house, then crossed a small empty lot. Passing by a barbed wire fence, he walked out to a field that had been plowed earlier that morning. He recalled that his father had just planted a new crop of sweet potatoes.

Peter looked up at the night sky and was surprised to see it filled with a spectacular array of beautiful lights, moving back and forth. His eyes were drawn to one very special light, which came slowly toward him. He stood very still as the ball of light got closer and closer. Peter's heart began to race with fear. The object stopped about eighteen feet from him. As he stared at the sphere, it suddenly began to spin and make a hissing sound. He stood frozen, an unseen force holding him. Fear overwhelmed every inch of his physical body, but he was finally able to overcome the strange force immobilizing him. He turned and ran back to the house. He heard the sound of an explosion and instantly felt a sharp pain in the back of his left hand. Blackness engulfed him. His next conscious recollection was his mother's warm arms holding him tightly.

Late one summer day in 1965, when he was seventeen, Peter and a close friend went on a fishing trip to one of the local lakes. (Our first abduction story involved a fishing trip as well, but Patricia's experience occurred four years later in a different state.) The trip was productive, and Peter and his friend wound up with their limit of fish. It was nighttime when they began their return trip home; Peter was driving, with George in the front seat next to him. The mood was relaxed as the radio blasted country-western music. Only a few miles from their destination, George peered upward through the windshield and said, "Pete, look at that airplane up there."

Peter slowed the car and looked up. "Hey, yeah, look at that!" They had arrived back in their own neighborhood by now, and Peter guided the car into the driveway and parked. They jumped out and immediately noticed a group of people standing in the street.

"Let's go over there and see what this is all about," Peter said. They broke into a run and approached the group. "What's happening, folks?"

A heavyset, gruff-looking man turned and pointed skyward. "Do you see that damn plane? It's been doing that kind of stuff for almost two hours now. I can't imagine what the hell is going on!"

Peter and his friend were excited. There actually was some weird kind of flying object cavorting above them. Peter turned to George and said, "Come on, let's grab the car and find a better place to watch."

They hopped into the car and headed for the open highway. George rolled down the side window and pushed his head outside so that he would not lose sight of the aircraft. Soon they found a good vantage point and pulled off the highway onto a dirt clearing. Peter jumped out of the car, his friend close behind. They ran to the front of the car and looked skyward. Both quickly realized this was no ordinary aircraft.

Peter reached into the car and got out a long black flashlight. It was the type capable of shining a focused beam. He adjusted the handle, pointed it toward the strange flying craft, and sent up a couple of quick flashes. He was amazed when he received the same series of flashes back!

He called out to his friend, "George, come here and take a look at this." His friend rushed over to see what was happening. "Watch this," Peter said. Pointing the flashlight up again, he began signaling. Again there were return signals from the craft. He flashed another round of signals, and suddenly the object began to descend. It came toward them slowly and stopped almost directly over them. Peter was excited but not afraid. He signaled with the flashlight, and the craft responded in a like fashion. Peter thought, "Whoever is controlling that ship is obviously intelligent."

Looking at the object above them, Peter and George were transfixed. They could see the underbody from where they stood. The craft was circular and appeared to be made of a shiny metallic substance. In the center was a large circle that contained three smaller circles.

Coming from each of these circles was a bluish-green glow. They observed a small opening in the exact center of the craft where the return flashing seemed to be coming from. This light was red in color. Peter began to doubt his sanity and had to turn his attention to something he knew was real. Looking up at the star-riddled sky, he noticed that his view of the stars overhead was blocked, and knew instantly that the craft above him was indeed a solid object.

The encounter lasted for about ten minutes. Peter was aware of the silence surrounding him as the object slowly began to drift off. It was a strange, eerie silence. The air did not stir.

From that moment on, Peter knew for sure that the object was not from this world.

Both teenagers, their curiosity still aroused, rushed to climb back in the car. Peter started the engine and accelerated out onto the highway. He was determined to follow the flying object for as long as he could. They noted that the craft was headed toward the city of Freeport.

Without warning, it turned 180 degrees and began to head back toward them. Again, they pulled off the road and stopped the car. The object paused for a brief moment and then, with a lightning burst of speed, disappeared into the night sky. Peter turned and looked at his friend, who sat motionless in the seat next to him. They both knew that they had just witnessed something they could neither laugh off, nor ever forget.

In 1971, Peter was living in the small town of Punto Filo, Venezuela. He had just been in an auto accident, injuring his left arm. The physician who examined him suggested that an X ray be taken of both his arm and hand. Peter agreed to this, and soon the doctor was holding a freshly taken X ray film in front of him.

The doctor asked, "Peter, did you ever injure your hand or have surgery performed on it?" Like Patricia, Peter told the doctor that, to his knowledge, there was no history of a hand injury, and he certainly hadn't ever had surgery on it.

Pointing to a small bright spot on the X ray, the physician said, "Do you see this, right here?" Peter nodded. The doctor told him that it looked like a surgical clip. He insisted that it was a solid metal piece of some kind, and asked if it caused any pain. Peter said no. The doctor, taking one last glance at the film, stated that it was probably nothing to be concerned about— probably an old cyst of some kind.

Still, the episode brought back an early childhood memory having to do with a strange night in a plowed sweet potato field. So Peter decided to test his hand with a stud finder, which is used in carpentry work, and found that the object seemed to project a magnetic field. (During the operation, we confirmed this with more sophisticated equipment.)

Three years later, in 1974, Peter visited his parents, who continued to live in rural Louisiana. His life had become quite settled and he had asked Laura, his girlfriend, to marry him. He made the announcement to his parents during the visit. Laura accompanied him and he was extremely happy and proud that his fiancée was so well accepted into the family. They had not set a date for the wedding yet, but planned to do so in the very near future.

After the visit, Peter and Laura got back into his car for the trip home. Laura gave him a peck on the cheek and asked, "Well, how do you think it went?"

Peter turned to her and said, "Honey, they loved you." He backed the car onto the street and started their journey home.

Almost immediately, Laura looked through the upper part of the windshield, pointed at the sky, and asked, "Peter, what is that thing up there?"

Peter slowed the car, leaned toward Laura, and peered at the sky. To his amazement, he saw a craft that looked much the same as the one he and his friend had seen in 1965. Immediately, he stopped the car. He was surprised to see the ship start to descend directly toward them. Laura's face turned ashen and she began to scream with terror. He reached out with both arms and held her tightly to him. "It's okay, darling, everything will be just fine," he whispered in her ear.

His mind raced—if only he could get some more witnesses. He pressed firmly on the horn, hoping this might rouse any nearby residents. He continued to pump the horn button as he rolled down the driver's window and started waving and pointing at the craft. Suddenly, it seemed as if the pilot of the strange craft understood the stress he was causing them. The object came to a dead halt, and then, with an incredible burst of speed, ascended and disappeared into the darkening sky. Peter has continued to have experiences, even after he was operated on to remove the implant from his hand.

THE FIRST SURGERIES
C H A P T E R   F I V E

I T was August 19, the day of the implant surgeries.
One by one the members of the team began to arrive.
In only a short time, the office was bustling with ac-
tivity.

I was pleased to see that everyone seemed to know
the job they were assigned and that they went about it
without much direction. Dr. A. was one of the first
people to arrive. The surgery room was carefully prepared for both surgery and videotaping. Television
cables extended from the room like a nest of snakes, all
heading in different directions. One destination was the
room adjacent to the operating room. This area was
equipped with a color television monitor and many
chairs. It was to be used as a viewing room for the
people we had chosen to witness the event.

Most of the preparations were completed by noon. The only thing we needed now were the patients. All of the people working with us that day were volunteers, and Derrel had made arrangements with someone who had offered to pick up our two surgical candidates from the airport and drive them to the office. They were due to arrive about 10 A.M.

Time passed slowly. Soon it was 1 P.M. and there was still no word from the driver or the patients. Derrel suggested we call the airline and ask if the flight had arrived on time. Janet called and found out that all of the flights that morning had arrived on schedule. We considered the possibility that there might have been an accident on the freeway that tied up traffic.

At 2:30 there was still no word on the whereabouts of our missing patients. The members of the surgery crew were becoming irritated and were mumbling about going home. I was frantic—I could see all my careful plans going down the drain because of one irresponsible volunteer. No one seemed to know how to contact him. People began to take turns nervously peering out the reception room window to look for approaching vehicles.

An hour later there were still no patients. At this point we began to seriously discuss whether to reschedule the surgery. I didn't know if I would be able to get everyone together again, and the very idea seemed exhausting. "Could you please just wait a little while longer?" I pleaded.

Suddenly I heard noise coming from the waiting room. "I think they're here!" someone said. We all immediately ran to the window to look out into the parking lot.

We saw Derrel greeting three people who were climbing out of a shabby old car. Aside from the driver, there was a woman who seemed to be in her early fifties and a heavily built man with an immense gray beard and clad in blue denim overalls.

A group of us gathered around the driver to ask about the delay. His reply almost started a riot. He said, "What delay? I gave these folks a nice tour of the beach area."

Derrel tried to be calm as he asked him, "Didn't you realize we were waiting for them? Their surgery was scheduled for this morning."

"I knew they were supposed to have surgery to-day," he retorted, "but you didn't tell me when."

At this point I interrupted them and said, "May I suggest we not waste any more time and get on with the surgery?"

The surgical candidates were escorted to the appropriate areas of the office to begin the processing procedure. A complete medical history was taken. Next, each underwent a thorough psychological examination that involved their abduction history. All paperwork, including the consent and release forms, were signed and placed in each patient's file. A chart was then prepared as a permanent record for each patient.

New X rays were taken and processed. My surgical nurse, Denise, withdrew the prescribed amount of whole blood for the transport medium which would hold the removed implants until they could be tested.

All the lab tests and pertinent data were immediately reviewed by Dr. A. and myself.

The small surgery room was filled with an eerie silence as our first patient, Patricia, got up onto the operating table. Her X rays showed that she had two objects in her left big toe, one on each side. I took my position sitting at the foot of the table; to my right was Dr. A. The camera operator was positioned so that he had a clear view of the surgery site. Denise would work from my left. Derrel was acting as the hypno-anesthesiologist, which placed him at the end of the table near Patricia's head.

The surgical prep had already been performed and the injection of a local anesthetic carefully administered. I applied an elastic tourniquet to the big toe. I looked at Denise and said, "Number ten scalpel blade," and she handed me the surgical knife. I suddenly realized that quite possibly history was about to be made. I took one final look at the X rays on the view screen, then pierced the skin, making the initial shallow incision.

A trickle of dark red blood began to ooze from the wound. Dr. A. responded by dabbing the area with a dry gauze sponge. "Number fifteen," I said. Denise placed the second surgical knife in my right hand. I began to deepen the wound, starting our fishing expedition in search of the mysterious metallic foreign body.

At this point Dr. A. placed the retractors carefully into the wound area. A forceful stream of bright red blood shot out and splashed against my sterile green gown. "Clamp!" I isolated the small bleeding artery and applied the clamp. The bleeding stopped instantly, and we were now free to carry on with the procedure.

"How is our patient doing, Derrel?" I asked. He responded with a hand signal indicating that everything was okay. The room filled with the constant drone of Derrel's voice as he continued to reinforce the hypnosis anesthesia. I was impressed with the effectiveness of the hypnotic anesthesia and convinced that the patient felt no discomfort. I administered a good healthy dose of my own concoction as well, which consisted of a long-acting anesthetic mixed with a quick-acting one.

Time passed quickly. We continued our search for the first implant by dissecting the soft and fatty tissue of the toe. Everybody's eyes were fixed on the open wound. Suddenly Dr. A. said, "What's that? I see something." Everyone held their breath. Was this finally the object? I cleaned the wound with a sponge and carefully inserted forceps to get a better look. I saw some tissue discoloration, but alas, no solid object.

Almost an hour had passed since the initial incision, and I could feel the tension mounting. Occasionally our photographer would enter the room and peer over my shoulder to try to get a firsthand glimpse of the proceedings. I asked Denise to reposition the X rays on the view screen so that we could take a fresh look. We did not have the kind of sophisticated equipment that would have allowed us to X-ray the toe on the spot, which would have helped us to find the object more quickly. The next time, I promised myself, I would have this equipment on hand.

The wound was retracted widely as I continued my probing efforts. All of a sudden the patient shrieked in pain and pulled her foot from our grasp. We were shocked—an anesthetized patient isn't supposed to do that! I reached up, grabbed her gyrating limb, and forcibly brought it back down to the table. Derrel immediately began to reinforce the trance state. I called for more local anesthetic and swiftly injected another large dose. It was only moments before things returned to the previous calm.

"What happened?" Derrel asked. I explained to him that the only time one ever saw anything like this was when the surgical procedure involved direct pulling on a nerve which was outside the field of anesthesia.

All at once, a crisp metallic click was heard. It broke the silence and reverberated in the small room. Everybody chimed in, "What was that?" I had touched something with my probe and it had made a noise. Again our patient objected with a violent movement. "Don't move," I gasped. "I think we've found something. Denise, pass me a Kelley clamp, quickly now!"

Everyone in the room remained still and silent. I carefully inserted the clamp into the wound, spread the jaws to their maximum, reached in, and grasped a solid object.

"Okay, nobody move now," I said. "Number fifteen blade, please. Thank you. Let's do it!"

With that, I began the careful dissection to free the foreign body from its fatty and fibrous tissue attachments. In just a few moments I announced, "Here it comes!" and with one final tug, the object was freed from the inner confines of the toe. The patient winced in pain and had to be given more anesthesia.

A shout of "Hooray" was heard from the viewing

room next door. We transferred the small object to a
white sterile gauze sponge. All eyes were fixed on the
peculiar object. Dr. A. peered intently at it and then
glanced at me. "What in the hell is that?" he asked.

I found myself staring at what appeared to be a
T-shaped mass that was dark gray in color and slightly
shiny. It looked fleshy, not metallic, and was about a
half centimeter long by a half centimeter wide.

The still photographer moved into position and
started to take pictures of the gauze sponge and its
strange contents. The flashing of the camera bulbs
added to the excitement. Soon it was our cameraman's
turn to record the object on video.

Dr. A. suggested that we find out what was inside
this exotic gray cocoon. He handed me a fresh surgical
knife. I clamped the specimen with a heavy-duty surgical clamp and began to gently incise the biological covering.

"What in heaven's name?" I exclaimed. "I can't
cut through the damn thing!" I asked for a fresh blade
and made a second attempt to cut into the object. "No,
it just doesn't want to give. Here, you try."

I passed both the object and the knife to Dr. A. and
assisted him in securing the specimen. He carefully began to stroke the tough gray membrane with the scalpel
blade.

"Roger, I'm just not doing any better with this than
you did. Let's put it aside for now and try again later."
With that, he handed me the sponge containing the
specimen and I passed it on to Denise. She placed the
object into the solution for transport.

Once the extracted specimen was put carefully to

bed, it was time to begin the second procedure. I asked Derrel if Patricia was prepared for the removal of the second object from her toe and he responded with an affirmative nod. For the moment, a quiet calm settled on the little surgical room as I replaced the constrictive band around the big toe. Dr. A and I again peered at the X rays on the view box. Then, with a fresh surgical knife I carefully made a superficial incision on the opposite side of the toe. Slowly the tissues began to separate. As I deepened the wound, small bits of fatty tissue were removed and placed in a regular specimen container. We knew that this object was going to be more difficult to find than the first one because of its position and smaller size. Time raced on, and I could tell that the surgical team was becoming tired and anxious. Our best efforts at finding this one had been of no avail so far. We removed the retractors from the wound and took a second look at the X rays. We decided that perhaps we were not quite in the right area.

I had been surgically removing foreign bodies for almost thirty-two years and was aware of most of the little helpful tricks surgeons have developed. I have found that by and large these techniques are no substitute for just plain skill and luck. The only devices which are truly helpful are sophisticated X ray instruments such as a Lixiscope or Fluoroscan, but unfortunately, these were not available to us on our limited budget.

After viewing the X rays a second time, we repositioned the limb and extended the incision toward the main body of the foot. With this new approach we were able to see a area which we previously could not. Only a few minutes passed before I spotted a small segment of abnormal, grayish tissue.

"What's that?" Dr. A. said. The team clustered around me, trying to get a peek into the wound. I gently inserted the probe and touched something solid. The patient winced in pain. "More local anesthetic," I told Denise. She passed me the syringe and I injected another two cc's.

Tension mounted in the room as I inserted a clamp and attempted to latch onto the object. A bead of sweat rolled from my brow onto my surgical mask. I was too close to my goal to stop in order for someone to sponge my forehead. I removed the empty clamp and asked Dr. A. to hold the retractors so that the wound was stretched to its maximum. Suddenly my clamp closed on something solid. With my left hand firmly holding the clamp, I inserted the surgical knife. With my right hand I meticulously loosened the soft tissue attachments holding the object.

With gentle upward pressure I was able to bring the object almost out of the wound. There seemed to be one final strand of tissue which needed coaxing. Dr. A., sensing the problem, reached for the dissection scissors and carefully cut the remaining attachment. A cheer went up from both the team and the adjoining viewing room. I tried to ignore the excitement as I placed the specimen on a clean sponge and began to measure it.

The cameras once again came to life. Because of the excitement, one of our team members got in the way of the video camera, resulting in a view of someone's back, just at the time the object was pulled from the wound. We didn't realize this until the tape was viewed long after the surgery was over.

The second object was much smaller than the first one, and looked like a cantaloupe seed with a few tiny tendrils dragging from its ends. We were amazed to see that this newly acquired specimen was also covered with a smooth, glistening, dark gray covering.

This time it was Dr. A.'s turn to attempt to open it. I passed him the surgical scalpel and the gauze sponge containing the specimen. We all stood and watched as he took the blade and gently tried to pierce the coating. We looked at each other in amazement. Once again, the object resisted all his attempts.

"Let's get on with the business of closing the wound," I said. I felt we needed to sew the wound closed and get the patient out of the operating room.

Soon the wounds were fully closed. The tourniquet was then removed and I watched as new blood flowed into the toe and the color changed from a rather pale white to a bright pink. Circulation was restored and the wounds were ready to be dressed. Derrel began to bring the patient out of the hypnotic trance. I asked how she was doing. Patricia herself piped up, "I'm doing just great. How are you doing?"

I told her that the surgery was over and everything was just fine. Once the patient was on her way to the recovery area, I quietly slipped out the door, took off my gloves, and headed for the doctors' lounge.

It felt good to have accomplished my first implant removals. I removed my soiled surgical gown and my mask and took a deep breath. Alice stuck her head in the doorway and offered me sandwiches, which were gratefully accepted. The amazing events I had just participated in flashed through my mind. It had been a long day, yet we were far from finished.

Dr. A. and I quickly made our way to the surgical room. Derrel was in the same position as he was during the first surgery, talking softly to the patient. I had decided that no hypno-anesthesia would be used this time. I wanted to see the differences between the recovery of patients who were hypnotized during surgery and those who were not.

Our second surgical patient, Peter, had replaced Patricia on the operating table. He was the large, bearded fellow, and he took up most of the space on the table. The table itself had to be rearranged due to the fact that the surgical area involved was Peter's left hand. As we entered the room I noticed that he shifted his eyes to meet mine.

"Hi, Peter, how are you doing so far?" I asked.

"Okay," he muttered under his breath.

The patient's left hand had been placed on a board, an extension added to the table for the purpose of supporting his arm. Peter's outstretched hand was placed palm down, and looked bright orange due to the spray applied during the surgical prep. This color extended from the fingertips up to the elbow. The skin had a rather strange appearance compared with the rest of him. Denise had shaved both his hand and arm. Dr. A. took a surgical magnifying loop and began to peer closely at the skin of the hand and arm areas.

"Rog, come over here and take a look," he said. I gazed through the loop, investigating every crack, pore, and orifice that I could view with the device. To the

naked eye, there was no evident portal of entry, no scar
or sign of any type of skin interruption that would have
allowed a foreign body to have entered the body. Yet it
was clearly visible on the X ray.

"Let's stop right here," I said. "Bert, could you
please bring in the gauss meter?"

Bert Clemens passed me the instrument. I took a
sterile towel and carefully wrapped the device so that
the meter could still be read. I placed it slightly above
the back of Peter's hand and pressed the on-off button.
Suddenly the room was filled with a strange, pulsating,
buzzing sound.

"What in the hell is that?" Dr. A. asked.

"I'll tell you what it is. This object in Peter's hand
is putting out one giant electromagnetic field," I replied. This was especially interesting, since we had
used the gauss meter on Patricia earlier with no results,
despite the fact that the objects we removed from her
toe seemed to be metallic.

At this point a couple of our observers suggested
that perhaps the device was being activated by all the
electronic equipment in the room. Mike Evans then
suggested that we take the patient out into the parking
lot, which is far away from any electronic influence, and
run the test again. I looked at Dr. A., and he nodded in
agreement. This meant we had to get the patient off the
table, take him outside, do the test, and bring him back
into surgery—which in turn meant that the prep and
positioning of the camera and cables would all have to
be done over.

I thought about it for a second and then decided.
"There is no choice; this is too important. Let's get on

with it. But let's be very careful and try to maintain those areas that are still sterile.''

With that, our group made its way from the surgical room, down the corridor, through the waiting area, and out the front door of the building. The parking lot was dark and only some isolated cars could be seen scattered about the area.

"Let's try it here. Bring over the instrument and let's see what she does,'' Dr. A. told us. I was still holding the gauss meter in the sterile towel as I brought it over to Peter's waiting hand. "Okay, Peter, just relax now. Let's have a look at the back of your hand again.''

Peter responded by holding his arm out straight. I moved the detection device into position and once again pushed the button. The still night was suddenly alive with a second chorus of that strange pulsating and buzzing.

Dr. A. said, "Let's get our patient back into surgery and get this thing out.'' We hurried back into the building to the operating room.

Peter was prepped for surgery again. Dr. A. picked up a fully loaded hypodermic syringe. The needle was just inches above the skin of Peter's left hand when the tough-looking patient suddenly lifted his head from the surgical table and exclaimed, "Damn, Doc, you don't have to poke me with that. Just go ahead and cut that thing out!'' I edged forward, looked our surgery patient squarely in the eye, and said, "Peter, just lie back and let the doctor do his job.'' There wasn't another word said, and soon the exposed part of the hand was numb with anesthesia.

Dr. A. started the surgery with a superficial incision

using a number 10 surgical blade. The wound welled with bright red blood. Performing as a second surgical nurse, I reached over and applied pressure with a sterile gauze sponge. Dr. A. was quick to continue the procedure, deepening the wound with a number 15 blade. Instinctively, my hand shot forth with another sponge. The wound was now bleeding profusely, and soon several small rivers of red liquid were flowing down Peter's arm and into the receptacle below.

Dr. A. placed two small retractors into the wound and with gentle pressure pulled the edges apart. With that, the process of fishing for the mysterious foreign body began.

The process was slow and tedious. We probed the wound, sponged the blood, looked at both X ray views, and peered deeply into the wound. An hour went by, and still the elusive object was nowhere to be found.

Dr. A. asked Mike to remove the radiographs from the view screen and bring them closer so that he might get a little different perspective. We gazed at the X ray films for several minutes and discussed the possibility that the hand might have been in a slightly different position during the time the X ray was taken. We finally came to the conclusion that the object was much deeper than was originally thought.

Dr. A. extended the incision slightly and increased the depth of the wound. The tendons glistened in the strong light of the operating lamp. He delicately retracted the tendons and exposed the deep tissues below.

All at once an audible click was heard. "I think I just touched something," Dr. A. said excitedly. I held my breath.

Peter suddenly piped up and said, "Hey, guys, that hurt!"

"Clamp," Dr. A. commanded without taking his eyes off the gaping wound. He once again entered the wound with the number 15 blade and in a fraction of a second pulled out the clamp, which had a small, dark-colored object sandwiched in its jaws.

With that, voices filled the operating room. "Is that it?" "Did you get it?" "What is that thing?" We could hear more questions coming from the viewing area. Mike turned the video camera toward the new specimen and adjusted the focus.

"Hold on, everybody," I said. "Let's take a look and see what we have here." The little gray object was then placed on the white sponge. I was shocked. It looked exactly like the object that I had removed from Patricia's toe just a few hours earlier. It was shaped like a small cantaloupe seed and appeared to be covered with a dark gray membrane. When Dr. A. picked up a knife with a sharp 15 blade and tried to cut the object away from the membrane that covered it, the same thing happened as before.

"I can't seem to cut into the damn thing. It's just like the other one." He handed the sponge containing the specimen to me and I passed it to Denise, who placed it in the waiting transport container. It took Dr. A. only a few minutes to suture the wound closed and apply the dressing.

The surgeries for that day were finally completed. Our visitors began to pour out into the hall. The mood was upbeat, and they showered us with compliments and affection.

But there was still much to be done. Not only did the patients require our immediate post-op attention, but it was also necessary to reevaluate them psychologically.

The day finally ended hours later, when the last person left the building and the dead bolt was latched on the outside door. I felt a tremendous sense of satisfaction and was eager to get on with final identification of the three mysterious objects we had removed.

TESTING
C H A P T E R   S I X

I RESTED on Sunday, mulling over recent events and taking time to discuss them with my wife. Sharon wisely advised me to move cautiously before jumping to any conclusions.

On Monday I was back at my usual routine, seeing patients. I finished up early, and began preparing the surgical specimens to be sent out for pathological testing. Derrel had taken the actual objects back to Texas with him. They were still in their fluid in the containers and still encased in their hard membranes; I kept the soft tissue specimens that had surrounded the objects in the patients' bodies. Since all the studies were to be double-blind, I had to decide what information I could include with the request form. I invented a fake research project dealing with soft tissue reactions in foreign body surgery. By using this as an excuse, I could ask the pathologists questions without their raising an eyebrow.

Since I was probably going to need all the help I could get, I made the decision to send one specimen at a time to the same laboratory. This way I could develop more of a personal relationship with the head pathologist.

I removed the specimens from the refrigerator in my office lab. The pathology lab forms were simple to fill out. The diagnosis was "soft tissue adjacent to foreign body." The first location was "big toe." I placed the specimen and the forms into a locked box, strapped it onto the front door of the office, and called the lab for a pickup.

I didn't put in a rush order—I wanted the lab technicians to take their time and do a thorough job. Meanwhile, Derrel tried to figure out a way to remove the gray membranes surrounding the objects so he could send the metallic portions to a laboratory.

About two weeks after the surgeries, Janet said there was a new batch of pathology reports on my desk to be reviewed before they could be filed in the patients' charts. This was one of those paperwork chores I had performed many times. As usual, I left it to the end of the day.

Most of the pathology reports were routine and pertained to bunionectomies, hammertoe corrections, and ingrown nail specimens. But suddenly I found myself reading a report labeled "soft tissue surrounding foreign body."

I sat up straight in the chair as I slowly read the

page. I had not expected the results from the first surgery so soon. As I read, I could not believe what I was seeing. There had to be some kind of mistake. These tissue samples showed no signs of inflammation.

It is impossible to have a foreign substance enter the human body without having the body react to that substance. Our system of defense is designed to ward off any invading substance, providing the body with the protection it needs. The result of this defense is inflammation.

Over a period of thirty years, I had removed many foreign bodies and looked at hundreds of pathology reports, and as far as I knew, the body reacted to all foreign bodies with inflammation. I considered the times I had implanted metal screws, plates, pins, and the like into the human foot. Some of these had to be reoperated on later and the metallic material removed, for a variety of reasons. In each and every case, the metal changed color due to mild oxidation and fibrous tissue formed around or adjacent to the screw heads. Also, when the specimen tissues came back from the path lab, reports indicated chronic inflammation.

The transplantation of donor organs is a prime example of how our bodies reject even the smallest differences in human tissue. If it were not for drugs which suppress the effects of our immune system, no transplant would be possible.

First thing next morning I called Chris, the pathologist who did the analysis. After giving her the case number and patient name, I got right to the point.

"I sent you a specimen of tissue I had removed that was adjacent to a foreign body, and you send me back a report telling me that the tissues are totally devoid of any inflammation. Come on, you know that's impossible.''

"Hey, I call them as I see them," Chris replied. "If I reported no inflammatory tissue, then by God, there wasn't any."

Another shocking finding was the large amount of nerve proprioceptors found within the tissue sample. There was no anatomical need for these specialized nerve cells to be clustered about a foreign body housed deep within the confines of a toe, next to a bone. Usually these tiny specialized nerve cells are only found in surface areas, such as the fingertips. They serve to conduct sensations such as pressure, temperature, or fine touch to the brain. Another area where proprioceptors are found is on the bottom of the foot. These serve to send messages through the spinal cord to the muscles, resulting in our ability to walk.

I couldn't imagine what purpose these nerves would serve deep within the soft tissues of the toe. However, they did explain the pain reactions that our patients had during their operations, despite the large amount of anesthesia they received.

I questioned Chris about these results. "Maybe you could give me a clue as to why you are reporting that this tissue is loaded with nerve proprioceptors. The stuff was taken from the deep, dark depths of a toe, just adjacent to the bone. How can you find nerve proprioceptors in that area?" I asked impatiently.

Chris explained carefully, "Look, Doc, all I do is slice up the tissue, apply my stains, pop it under a microscope, and report what I see. I really don't care if your specimen came from the moon. I'm sorry, I just don't have the answers for you.''

Recently scientists at UCLA have been working with a device which converts the bioenergy of the human body into the type of electrical energy that can be used to drive motors in artificial limbs. Perhaps if these nerves were in deep tissue, they could be tapped and used to generate the energy to run some sort of implanted object.

I decided to prepare the next tissue sample, arranged for a pickup by the lab, and called Chris the next day to alert her so she could intercept the specimen and perform the analysis personally.

Her question took me by surprise. ''Roger, what in heaven's name are you up to with these specimens?''

I had to think of something fast. If I told her the truth, it would ruin the double-blind study.

''It's no big deal. I'm just doing a little research project on foreign body reactions. By the way, I need to keep this kind of quiet. You understand, don't you?'' She seemed satisfied, and agreed to put a rush on the new sample when I told her I needed the results for a paper I was writing.

A few days later, I noticed another neat pile of lab slips on my desk. I sat down and quickly went through them. There it was! I hurriedly opened the envelope and began reading the document. I was dumbfounded. The report was almost identical to the first one. Nerve cells and no inflammation.

''This is too much,'' I blurted out loud to an empty office. Although it was 6:30 P.M., I tried calling Chris at the lab. Fortunately, she was still there.

"Hi, Chris, this is Roger. I want to thank you for getting me the results back so soon. Also, could you verify the fact that you did this sample yourself?" I asked.

"Yes," she answered, "I was the one who did it, just as I promised you the last time we talked."

"Of course, you realize the results of this one are almost exactly the same as the previous specimen."

"Well, now that you mention it, I guess they are pretty similar. But what's the big deal? You said they were from a foreign body reaction, and that's what you sent the first time, wasn't it?"

"That's true," I explained, "but this one was from a different foreign body and taken from a different place."

"Wow! That is a little strange, but there's probably some logical explanation."

For the next forty-five minutes, she presented one idea after another, all of which I countered. Finally I asked if she would photocopy and send me information from the latest *Robbins Textbook of Pathology*—which is the bible of pathologists—referring to inflammation and soft tissue foreign body reactions. She agreed.

I reasoned that the next specimen should be done by someone else in order to get a fresh opinion. I didn't call Chris about this one, and made a written request on the lab order form that Chris not perform the analysis. I knew there were many other capable pathologists in the lab.

It was time for some serious research. I logged onto the Internet, searching for information on foreign body reactions in the human body. My search took me to medical libraries at Harvard and Stanford Universities. Then I looked through my own books, some of which were outdated. My findings proved to me that the human body had not changed since the time I graduated from medical school. I saw no reason for the lack of inflammatory response found in the microscopic analysis of my two specimens. Perhaps, I thought, the third specimen would prove different.

It was almost a week before the results came in on the next specimen. By the time they arrived, I was ready to be confronted by the same mystifying analysis. I was not disappointed. The new report was almost exactly the same as the previous two. Only the language was slightly different, because the analysis was performed by a different pathologist.

I decided to call Derrel and check on his progress. "Did you get the membranes off the objects—and what about the metal inside?" I asked.

"Before trying to remove the membrane, I subjected all the objects to black ultraviolet light, and they all fluoresced a brilliant green color. This is the same color we've found on abductee victims following an experience. Do you think this is one hell of a coincidence or what?"

I remembered one such case I'd personally helped investigate, a lady who came to my office during one of Derrel's visits to California, shortly before the surgeries. We both examined her. I was interested in the medical aspects of the case and Derrel asked about her abduction experience. We took her into an examination room and dimmed the lights, and I used the ultraviolet black light to examine portions of her skin. Suddenly I

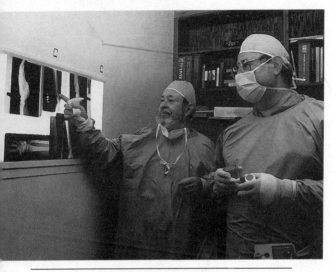

Dr. Leir (*left*) indicates the location of an implant on
X ray to an assisting surgeon.

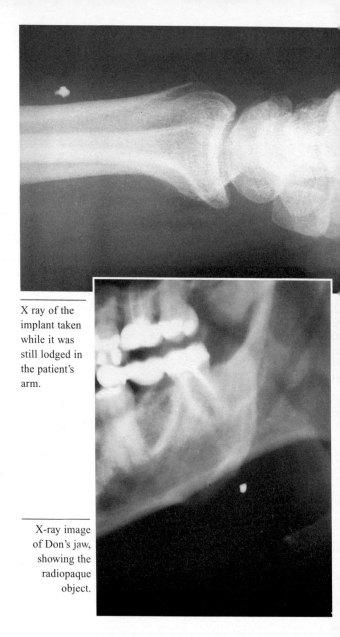

X ray of the implant taken while it was still lodged in the patient's arm.

X-ray image of Don's jaw, showing the radiopaque object.

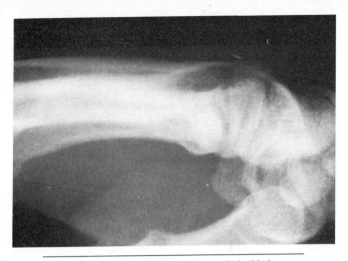

X ray of Paul's hand, showing an imbedded
radiopaque object.

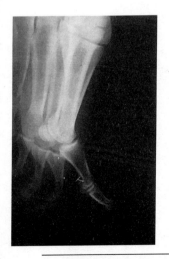

X-ray image of Patricia's foot, showing two
radiopaque objects.

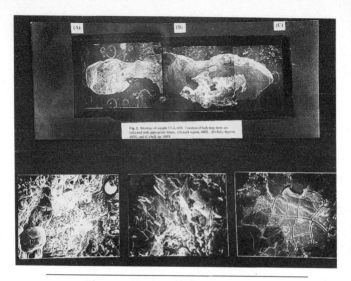

Electron photomicrographs of sample T3
from Patricia's foot.

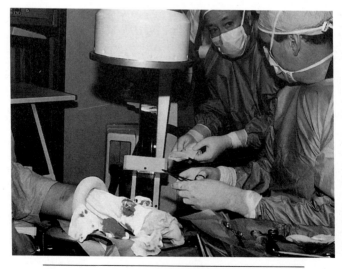

Removing the implant from the patient's arm, in an area
near the wrist.

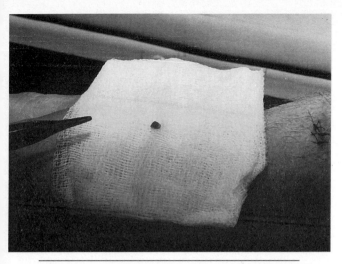

The implant as it appeared moments after removal, still encased in a membrane composed of body surface tissue.

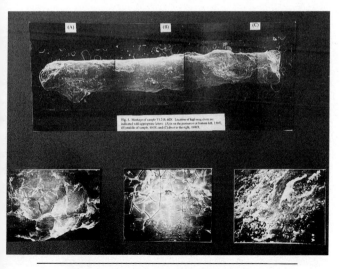

Electron photomicrographs of sample T1, 2 from Patricia's foot.

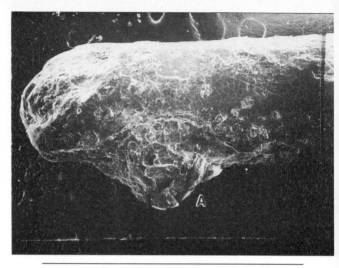

Close-up of left end of sample T1, 2 from Patricia's foot.

Close-up of center of sample T1, 2 from Patricia's foot.

Close-up of right end of sample T1, 2 from Patricia's foot.

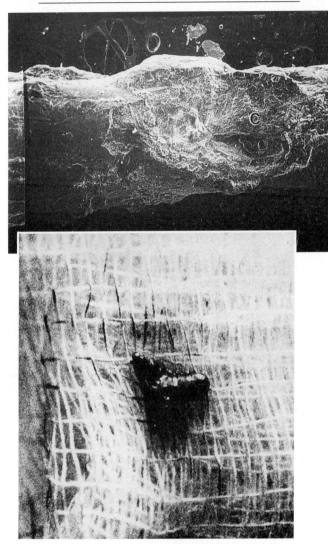

Patricia's T-shaped extracted object, covered
with tough membrane.

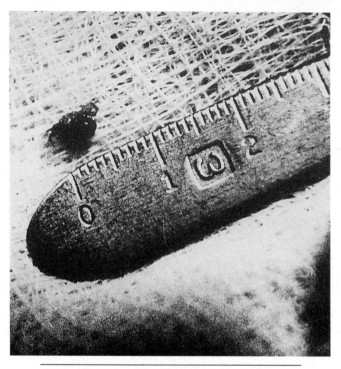

A cantaloupe-seedlike object from Patricia's toe.

observed a pink fluorescence in the palms of her hands. I immediately called Derrel into the room and asked if he had seen that particular color before. Much to my surprise, he stated he had seen the very same color on the palms of other abductees.

I asked the patient if she had touched anything with her hands which could account for the pink stain. She told me that perhaps she had gotten it from the steering wheel of her car. I decided I would try to remove it from her skin with alcohol. I picked up a gauze sponge, wet it with alcohol, and scrubbed her palms. Again I dimmed the light and turned on the ultraviolet. The pink stain was gone. I said, "I guess this was something she must have picked up from an ordinary source—otherwise it would not have washed off."

Derrel agreed with my conclusion. Most of the stains he had seen could not be washed off with any common substance. He told me that in the cases he had studied, the chemical producing this type of stain was subdermal, well below the surface of the skin. The patient was then taken back into the consulting room, where Derrel began to ask her questions about her abduction experiences. He watched her carefully while she answered, and made notes of her smallest gestures and reactions. I left them in the examination room while I worked on some office chores. By the time I got back, almost two hours had passed. He appeared to be finished with the exam and the patient was preparing to leave.

On the spur of the moment I turned to the patient and said, "Before you leave, could you just give me a moment? I would like to check your hands one more time.'' Once again I took the patient into the examining room, dimmed the room light, and turned on the ultra-violet. I asked her to hold up her palms and keep her fingers outstretched. I moved the black light over the area and, much to my surprise, her palms were once again glowing a brilliant pink. Perhaps it was true that the substance had been subdermal; maybe I had wiped it off the superficial skin and it had now had time to seep back through again. I did not see any other way the fluorescence could have returned.

Now, Derrel went on to tell me he was able to remove the membranes from the metallic portion within by drying the specimens. The metal portions were dark gray in color and had an irregular surface. Also, they were highly magnetic and would stick to any ferrous metal object. He was waiting to hear from his Ph.D. friend at the University of Houston, who had agreed to do the metallurgical testing. I asked him to send me portions of the membrane so I could send them out for analysis.

Almost a week later, Derrel's membrane samples arrived by Federal Express. There were four plastic containers of small dark particles. I gathered up the vials, headed for the office laboratory, and placed them carefully in the refrigerator.

When I prepared the pieces of membrane for the laboratories, I made sure to keep the different mem-branes separate from each other, so that each portion could be easily related to the metallic object it had encased.

About a week went by before the lab reports began to come in. I was again astounded to see the results of the analysis. The membranes were composed of just three biological elements: a protein coagulum, hemosiderin, and keratin. Protein coagulum is derived from clotted blood and consists of pure protein. Hemosiderin is an oxygen-carrying iron pigment that is closely related to hemoglobin. Hemoglobin is found in our red cells and binds with oxygen in the lungs. It is circulated to all portions of the body's tissues, where it gives up its oxygen and takes on carbon dioxide. It is then transported back to the lungs for expiration, and the process begins over again.

Keratin comprises the outermost covering of our bodies. It is the most superficial layer of the skin. Also found in our fingernails, toenails, and hair, it is probably the most cared-for substance in our bodies. About 90 percent of the cosmetics industry depends on our treatment of this outer keratin layer.

I asked myself how this strange concoction of natural biological material could have come about.

My next step was to consult the pathologist who had performed the analysis. Our conversation went as follows: "Hello, this is Dr. Wong. Did you have some questions on the sample I reported on?"

"Yes, I'm the surgeon who removed this sample and I had a few questions about your report."

"Sure, go ahead and ask."

"To start with, how do you think the protein coagulum was formed?" I asked. There was a moment of silence.

"Most probably there was some internal bleeding which became organized over the past weeks. What we are seeing is the remaining portion of the organized clot that has not been carried away by the body." He rattled this off in a single breath.

I wasn't convinced. "The only problem I have with that is, this object has been in the patient's body for forty-one years. Don't you think that after this long the body would have had enough time to remove this debris or wall it off with fibrous tissue?"

Again there was dead silence, and finally he began to speak again. "Well, Dr. Leir, I didn't know those facts, but you know the human body is not always predictable."

I thought, "What kind of a line is he handing me? Certainly nothing helpful."

"Dr. Wong, could you possibly help me understand how the keratin could be part of this, since the specimen was removed from the deep tissues adjacent to the bone?" There is normally no keratin in that part of the body.

His answer was quick. "Most probably, you dragged in some skin when you made your initial incision and this became involved with the deeper mass."

"Dr. Wong, let me make this very apparent. I am a very careful surgeon and I can assure you I did not drag superficial tissue into the depths of the wound. In addition, even if that was done, how would it be possible for the keratin strands to actually become part of the tissue we are analyzing? Since the object had been in the body for many years, I would think that the structure of this organic tissue would be stable and not allow for penetration of substances like keratin."

"Perhaps the original penetration brought keratin into the wound." He was referring to the wound that should have been made when the object originally entered the body.

"Dr. Wong, that is the most plausible answer. However, there is one small problem. There is absolutely no evidence of an obvious penetration wound. We searched the extremity with a loupe and could not find even a hair or pore that appeared out of place."

"Well, my friend, as I told you a moment ago, the body works in very mysterious ways, and sometimes we just don't understand enough to come up with all the answers." With that our conversation came to an abrupt end.

This was typical of the way my conversations went with other pathologists in reference to this subject. In considering all the factors pertaining to these membranes, I reluctantly came to some far-out conclusions. We knew these objects did not cause the body to react to them with an inflammatory or rejection response. The reason for this might have to do with the membrane.

I discussed this possibility with a couple of bioscientists and they concluded that, if this was true, we had just discovered something that could revolutionize medical science. If this membrane could be duplicated, we could use it to wrap anything we wanted to instill into the body and the body would not reject or react to it. Even transplant organs could be wrapped in this substance, with the result that patients would not have to take antirejection medication for the rest of their lives. This is an avenue of research we are pursuing at this time.

There was something specific about the keratin

which I couldn't get out of my mind. Remembering that I could not cut through the membrane with the surgical scalpel, I began to wonder if the keratin was responsible for its tensile strength. I began to think about how such a strange brew of ingredients could be produced, and suddenly a light turned on in my brain. I remembered that one of the most common marks found on the bodies of alleged abductees was the "scoop mark," which looked as if someone had taken a tiny teaspoon and scooped away some superficial skin. I had seen many of these during the last few years.

They healed in a peculiar fashion. Most superficial abrasions usually heal in a week and form a scab that later peels off, leaving only a red mark on the underlying skin. These scoop marks did not heal in that way. They had a very shiny base which appeared moist, but if the lesion was touched there was no moisture. This condition lasted for just a few days, after which time only a perfectly formed depression was left in the skin.

I began to wonder if there was a device that could scoop keratin from the surface skin. Perhaps keratin obtained in this manner could be incorporated into a gel-like mixture which would then form a membrane.

I made copies of the reports, filed the originals in the patients' charts, and sent the copies off to Derrel. Almost a full month went by before I talked with him again. I wanted to know what progress he was making with the analysis of the metal portions. I was beginning to receive phone calls asking me what the objects were made of, and I had to put the callers off. When we spoke, he explained that there had been no progress on the metal analysis because the Ph.D. program at the university was on temporary hiatus. It would be at least another month until it was up and going again.

I suggested that we look elsewhere to have the tests performed. Derrel reminded me he was having these tests performed without charge and asked if I knew where we were going to get the funds to pay for them ourselves. I told him that I had a wide circle of friends and would look into it.

I spent the next few days making phone calls to people I thought could help. I also asked Alice to look over the MUFON membership list to see if anyone in our group might be in a position to help us analyze the implants. About two days later I received a call from a friend of mine named Jim, who gave me the name and number of a friend of his, John, who he said could probably help. We set up a meeting for the three of us in Ventura the following week, about a half hour's drive from my office.

The lunch went well. John had a background in mechanics and engineering. He also confided in me that there was a possibility both he and his brother had had an abduction experience many years ago. The more I talked with him, the more I realized he had some of the psychological traits of the typical abductee.

John said his brother used to work for a very large laboratory and that he would ask him if he could get some of the tests done on our samples. There would be no charge because his brother was still owed a lot of favors from his friends at the lab. This was wonderful news, and I was extremely excited. I told him I would be in touch with Derrel and that we would get him the material samples.

After I arrived back at my office, I immediately
called Derrel and gave him the news. He would have to
bring the samples to California as soon as possible, I
told him. Within a few days, Derrel arrived and we
handed the samples over to John. I began to believe we
were on the threshold of a major finding.

Derrel went back to Houston and I continued with
my everyday life, always hoping to hear news about the
implant testing results. I attended several MUFON
meetings, and at each one I was asked to reveal the
results of the implant research. I felt an obligation to
our group to keep them informed and to explain that
because of the high cost of getting tests done, we had to
rely on favors, which meant it would take a longer time
to get the results. Several individuals suggested names
of people who might be able to help.

Over the next couple of weeks I called everyone
who was recommended. While some of them had either
changed jobs or could not do the tests we needed done,
others were quite congenial and left the door open. I
called Derrel and suggested he take more samples of
the materials and bring them to California for testing by
another source.

The following week we turned over another set of
samples to Brian, who was recommended by one of the
MUFON members. He told us he would have the results back in two weeks. That sounded great, and we
were happy with our decision.

I happened to get a call from John while Derrel was
in California, so we set up another meeting. I was ea-
gerly anticipating the results of his brother's analysis.
Derrel and I drove to Ventura again to meet with him.

John said he had some preliminary results, and I took out my pad and prepared to take notes. He told us that his brother's testing revealed that one of the compounds contained in the sample was called boron nitride. John knew that this substance was found in nature, but was not sure whether or not it was also manufactured. Many so-called natural substances are made in laboratories rather than extracted, because it's cheaper to produce them that way. If boron nitride was one of these, it meant that the object could have been manufactured. He told us he would call his brother to find out. He also mentioned the presence of iron and magnesium.

I asked him if he knew what the ratios were or what type of tests had been done, but he was unable to answer those questions. He promised us that the final analysis would be done as soon as possible.

Several more weeks went by, and still there was no word from either of the people who were supposed to be performing the tests.

I called Derrel and asked his advice. He said he would call both individuals to see what he could find out. The following day, my secretary told me John was on the line and wanted to talk with me right away. I was excited because I assumed he was going to tell me he had the final analysis.

"Roger, I heard from my brother and he told me that boron nitride is not found in nature and is some sort of a high-tech compound."

"Hey, that's really great. What else did he find?" I asked.

"What do you mean, what else?" John replied. "I

was happy that he provided this much information. What else do you expect to find?''

I began losing my patience. I really didn't want to antagonize John, since he was doing us a favor, but the project was too important to let this go on any longer. I asked, ''John, when do you think you can get us a written report?''

''Are you crazy? My brother is trying to do us a favor by having someone run these tests on laboratory equipment without them finding out about it so it won't generate a charge. You can't expect them to write a report.''

I was stunned. All this time we had been sitting and waiting, and this was the result. What was I going to tell people who called and asked for information?

I finally came to the conclusion that back-door research was not going to work.

LAS VEGAS
C H A P T E R   S E V E N

WE had to find a way to get funding for legitimate research so we could have these tests done by a reputable laboratory.

I called my cousin Ken, who is a university professor, to explain our situation and see if he had any suggestions. I told him about the problems we were experiencing with the testing. He supplied a few names of people who might be able to help me.

One was Dr. John Alexander, a scientist Ken had worked with in the field of near death experience. He thought Alexander now lived in Las Vegas and was working with a research company. When I called the number Ken had, the company's name was given out as the National Institute for Discovery Science.

When Dr. Alexander came on the line, I introduced

myself as a cousin of Dr. Kenneth Ring and told him about our research. He appeared to be very interested. I was surprised to find he already knew about the surgeries I had performed and also knew who Derrel was. When I asked if there was any way he could help solve our problem, he requested that I fax him any data I could, and said he'd get back to me quickly. I compiled the material and faxed it to him right away.

The next morning, my telephone at home rang at 7 A.M. I was sound asleep when Sharon shook me and handed me the portable phone. I couldn't imagine who would be calling so early.

"This is John Alexander. I received the information you faxed me last night and I have Bob Bigelow on conference with us."

The next voice I heard was that of Robert Bigelow. "Dr. Leir, do you mind if I call you Roger?" he asked. "Just go ahead and call me Bob."

" 'Roger' will be just fine," I said, trying to collect my thoughts.

Alexander explained the aims of the National Institute for Discovery Science (N.I.D.S.). The organization had been set up by Bigelow to serve a dual research purpose. One line of research was into the nature of the mind; the other was the study of unusual aerial phenomena—namely, UFOs. Alexander was the organization's director and Bigelow was its CEO. Alexander expressed deep interest in the project we were working on and said he was impressed by the biological data. He and Bigelow spent almost an hour asking me questions about the project, and when they were done we agreed to get in contact again shortly. They were going to present the material to several members of their board.

Two days later my secretary advised me that there was a Mr. Bigelow on the phone, requesting to speak to me immediately. I was shocked that he was calling me so soon. I excused myself from the patient I was involved with and made a dash for the phone.

Bigelow invited Derrel and me to visit the N.I.D.S. offices in Las Vegas with the specimens. I told him I would get in touch with Derrel immediately and arrange for a time.

I ran back to the treatment room and finished with the patient. Then I called Janet and asked her to tell my other patients I was running a little late. I dialed Derrel's number several times, and each time it was busy. In desperation I called the operator and asked her to break in for an emergency call. When I finally reached him, I got right to the point. Derrel consulted his calendar and told me what dates he was free. We arranged to meet a week later at the Las Vegas airport and share a cab to the N.I.D.S. offices, then return the same evening.

The following week, I met up with Derrel at the Las Vegas airport and we took a taxi to N.I.D.S. Mr. Bigelow's office building was surrounded by a stone wall about six feet in height. As our driver pulled into the driveway, an armed guard appeared out of nowhere and approached the cab. Once he confirmed that we were in the right place, Derrel and I got out and proceeded to a very strange-looking structure which seemed like a cross between an apartment house and an office building. At what appeared to be a reception desk, we were asked to have a seat and told that Mr. Bigelow would be right down.

We sat in chairs that blended in with the decor of the room. This area of the building was paneled with the most beautiful wood I have ever seen. Every piece of furniture appeared to have been specially made for the room. There was not a speck of dust to be found and every item was in its proper place. It seemed we were involved with someone of impeccable taste.

After about ten minutes, the secretary rose from her desk and started up a curved flight of stairs constructed of the same fine wood that was used throughout the office. In a few moments she returned with two men dressed in sport coats and ties. The gray-haired man introduced himself as John Alexander. We exchanged greetings and he introduced us to his companion, Robert Bigelow. I was a bit surprised at Bigelow's appearance. I had pictured him as a much older man. Although handsome and distinguished, he looked young, maybe in his late thirties. I would soon learn, however, that he had grandchildren.

Alexander invited us to join them for lunch, after which we'd have our meeting. During the meal, we made small talk about how Las Vegas had grown over the years. I asked Bigelow how long he had lived in the area and he was quick to say that he was born there. I thought it better not to ask him in what year.

The ride back to N.I.D.S. took only a short time. Derrel and I soon found ourselves sitting in Bigelow's private office. It was a spectacular sight. The decor was magnificent, and we were surrounded by carefully chosen art objects.

When we got down to business, Alexander led the conversation. He reiterated that N.I.D.S. was formed for research in two distinct areas of interest: investigation of the mind and study of aerial phenomena. Both men explained the plans for future development of the organization. Derrel and I each stated what we hoped to obtain from N.I.D.S.

Bigelow told us that he was impressed with our research and would present our findings to the board, which was composed of sixteen of the nation's top scientific minds. Alexander then offered to drive us to the airport. We exchanged good-byes with Bigelow, gathered our belongings, and followed Alexander to the car. The trip to the airport was enlightening. Alexander explained how he became involved with Bigelow and talked about experiments in remote viewing and psychokinesis, and also about his participation in the development of the country's nonlethal weapons program. He proved to be a fascinating person. He also related his involvement with my cousin Ken and told us of his continuing interest in the study of life after death. There were many times when Derrel and I looked at each other in amazement.

On Monday morning of the following week, I was once again awakened out of a sound slumber by the jarring ring of the telephone. I picked it up groggily.

"Roger, this is John Alexander. I have Bob on the other line. We would like to take a few minutes of your time, if that's okay."

"My gosh," I thought, "both of them at once! What could they possibly want at this time of the morning?" It didn't take long for them to get to the point.

This time it was Bigelow who carried most of the conversation. He told me he had discussed some of our biological findings with several members of the board and they were interested in helping us. In fact, he had one of them standing by, and asked my permission to put him on the line. I agreed, and soon we were having a four-way conversation.

This board member was proficient in two areas. He had a medical degree and was a board-certified neurologist. He also held a Ph.D. in physics. I was impressed, to say the least. Our conversation became more complex, with a great deal of medical jargon that I am sure was over the heads of Bigelow and Alexander, who stayed silently on the line. His questions were not antagonistic and gave me the impression he was sincerely interested in my answers.

The next part of the conversation took place among the other three as I sat quietly listening. It seemed they thought it would be a good idea to present the material to the entire board, and they asked me when we would be free to come back to Las Vegas. John reminded his colleagues that the board was scheduled to meet on a weekend in about two weeks. He asked if this was too short notice. I told him I would call Derrel and we would rearrange our schedules accordingly.

A package soon arrived containing my itinerary and plane tickets to Las Vegas, and I got a phone call from Derrel telling me that he had received the same package. The flights would arrive in Las Vegas at approximately the same time. I also received a list of items I was to bring with me. Derrel told me that his package contained a different list, one of the most important things on it being the objects we had surgically removed.

Two weeks later, Derrel and I again found ourselves being escorted into the N.I.D.S. building by two armed guards. This time we entered what appeared to be a rather lavish dining area, filled with many tables, all covered with white linen tablecloths. The silverware was displayed elegantly on cloth napkins. There were a few individuals standing around the room eating hors d'oeuvres. We were taken into a kitchen area and invited to help ourselves to any of the food that was displayed. One quick glance around the room aroused my appetite. There were platters containing every imaginable type of cuisine. The bread-and-roll section looked like a miniature bakery. There were glass bowls containing all kinds of salads, as well as an array of steam trays with heat and steam billowing up from around their lids. Derrel and I began piling food onto plates. Then we found a table set for six and began to eat.

During the next ten minutes, the room filled with people. The background conversation hummed with the tone of scientific talk. We deduced that the board was taking their lunch break. Suddenly I heard a familiar voice across the room calling my name. It was Bigelow, surrounded by several individuals I didn't know. He gestured for us to join them, and Derrel and I made our way over. Bigelow excused himself from the crowd, stuck out his hand, and gave us the kind of greeting one would give old chums. He then introduced us to members of the board and their guests. It seemed that this meeting was to be more than just an ordinary board gathering. One of the people we were introduced to was Jacques Vallee. Both Derrel and I had read many of this famous ufologist's books, and we considered it an honor to make his acquaintance. Bigelow told us to go back to our table and enjoy our lunches; he would come over and talk with us about the procedural issues of the meeting shortly.

Just as we were about to have dessert, Bigelow approached our table and sat down. He told us there was a schedule and the format would be adhered to rather strictly. The appointed time for our presentation was 2 P.M. He advised us to take a stroll through the grounds while we waited. At the proper time he would send someone to escort us to the meeting room.

Derrel and I went outside onto the grounds of the magnificent estate. We found a path that took us through a beautiful garden filled with the aroma of roses and other flowers. We walked through canyons, mountains, and bridges that crossed streams of flowing water—all artificial, of course. Soon we found ourselves deep within a darkened cave.

Once inside, I reached up and with both hands began to turn a suspicious-looking rock. All at once, a huge boulder began to slide out of the way, revealing a dimly lit secret chamber. There were several places to sit in it, on what appeared to be seats carved out of the rocks. I guessed that Bigelow had built this secret space as a getaway from the everyday world of business, finance, and family problems.

Soon we were out of the cave and back in the warm

sunlight. We wound through artificial forests and miniature mountain areas and came to another chain-link
gate. We opened the gate and entered an area that reminded me of Disneyland. It had the same artificially
created beauty, and everywhere we looked was some
sort of a fun ride or game that children could play. I
guessed that Bob loved his grandchildren and had built
this area especially for them. Then I heard a voice saying, "They are ready for you now."

A guard appeared to escort us to the boardroom. He
led the way up a spiral staircase and through an area I
recognized from our previous visit. We were allowed to
take a short detour to claim the items we needed for our
presentation, and soon we were entering the inner sanc-
tum of the Bigelow domain.

The boardroom was typical of what you might expect at a corporate giant such as General Motors; I
would not have expected to find anything like it in Las
Vegas. The room was huge and constructed of the same
fine wood seen throughout the building. As I entered, I
caught sight of a huge television screen. It was the type
of monitor you see in a major-league stadium, composed of individual picture tubes combined to form one
giant image. Hanging next to this was a pull-down
screen. The conference table was in the shape of a
horseshoe. There was enough space around the table to
accommodate not only board members but any guests
as well.

John Alexander made his way through the gathering and greeted us. He told us I was to do the entire
presentation and advised Derrel to take a seat and to be

ready to answer questions. My heart fell to the floor
when I heard this. I was a little-known foot surgeon
from a small southern California community. I was nervous about talking to some of the foremost scientists in
America.

I heard the door close with a dull thud, and instantly the room became silent. Bigelow spoke in a soft,
authoritative voice. He directed the board's attention to
me and made the introduction. He told them I would be
presenting the material they had previously discussed
and that I would show slides and videotape. Afterward,
they would be allowed to ask me questions.

Bigelow signaled me to begin my presentation. I
felt petrified with fear. Yet there was nothing to do but
go on, so I plunged in, telling everyone that I would
pass out a prepared document. I reached into one of our
briefcases and pulled out a bunch of them. Derrel rose
from his seat and helped to distribute them to the board.
After the shuffling of papers had stopped, I introduced
our videotape of the surgeries.

When the video presentation was over and the
lights came back on, the questions began. They were
all of a technical nature and I answered them to the
best of my ability. As the questioning continued, I noticed that the board seemed to be coming around to
my side. Some of them began to answer the questions
for me. At this point, I knew I had made a favorable
impression.

From then on it was smooth sailing. Whenever I
was asked a question I could not answer, I would
merely glance at one of my supporters and he would

jump in to answer the question for me. The more I talked, the more confident I became.

After the presentation, we gathered up our material and walked to an adjoining room filled with comfortable chairs and couches and a number of ornate tables scattered around. I was surprised when most of the members of the board began to walk over to Derrel. They asked if they could see the collection of specimens that alleged abductees had given to him over the years. He was quick to oblige, and soon there was a crowd standing around him. Derrel was in seventh heaven. He carefully explained the origin of each specimen, along with the details of the particular abduction experience involved. I found this quite fascinating, since scientific types, I had often been warned, had no interest in hearing tales of alien abduction.

The board was about to convene again. One of the guards was sent to escort us back to the dining room where we had started the afternoon. We were invited to relax and make ourselves comfortable and told that someone would be coming down to meet with us periodically. Derrel and I sat and discussed the day's events. It was his opinion that we had done very well. I wished I could have agreed with him. I felt the outcome was a total unknown.

We took the time to review the situation, wondering what they would offer, if anything. It wasn't long before our questions started to be answered. Bigelow appeared in the doorway, walked over to us, and pulled up a chair. He told us that in his opinion, I had made an excellent presentation. The board would be voting shortly on their decision, he informed us, and he would be popping down from time to time to keep us up to date on the situation.

Derrel and I sat looking at each other. A smile crept over his face and he said quietly, "Roger, I think we did it. Isn't it great?" I agreed with him wholeheartedly.

Bigelow's trips up and down the stairs came at regular intervals. Each time he arrived, he would bring us additional information. Finally, at about 5:30 P.M., he came down and asked if we were interested in taking a walk. We jumped at the chance to get some fresh air— and, we hoped, some good news. Bob led us into an area of his parklike garden. As we strolled along, he informed us that the N.I.D.S. board had voted to help us in our research efforts. We were elated. With a handshake, we agreed to turn over our surgical specimens for analysis.

There were certain stipulations to the agreement. They wanted us to write a scientific paper, to be published in a world-class journal. We were delighted with that idea. For our part, we insisted that the data uncovered by the testing had to be revealed to the public. This was our only stipulation, and we were adamant on this point—without it, there could be no agreement. We agreed to allow N.I.D.S. to select the laboratories of their choice. Our working relationship had to be one of mutual trust.

Derrel and I sat down once more in the dining room. We gathered up the specimens and began to code them. All tests were to be double-blind. Only Derrel and I would keep the key which told where each object originated. We then gave Bigelow the specimens and shook hands. I am sure that he was as excited as we were.

We said our good-byes, gathered up our materials, and were escorted to the car which would take us to the airport for our return flights home.

THE SECOND SURGERIES
C H A P T E R   E I G H T

THE first request N.I.D.S. made of me was to send them a detailed budget for more surgeries and related research. I already had three potential patients lined up.

We had received many applications from people claiming they had implants they wanted removed. From this list I chose three who I believed met the criteria— one male and two females. All of them believed they had been involved in an abduction experience. The only slight variation was provided by the male candidate, who was not completely sure whether he had been involved with aliens or the U.S. military.

We knew from past experience that some alleged abductees claimed military involvement, but this assertion was always difficult to prove. Don was an employee

of the government who held a low-level clearance. He
was having problems with a painful tooth, and asked
some of his coworkers for the name of a dentist. One of
his bosses advised him to try the dentist across the
street from the office building where they worked.

After he had dental work done there, he began
hearing voices in his head. He blamed this on his trip to
the dentist, and went back and complained, but was told
that the dental work had nothing to do with the voices.

When we examined Don, he had a metallic foreign
object in his left lower jaw; the object was easily seen
on a radiographic examination. He swore he didn't
know how it had gotten there. He related the voices he
heard to this object. We told him that after the surgery
was performed and the object removed he might continue to hear the voice communication. He understood
this but nevertheless wanted the object removed. Despite hearing voices, he passed the psychological tests
we gave him.

The second surgical candidate was a female with an
object in the front portion of her leg, adjacent to the
shinbone. X rays showed only a small round shadow in
this area. In addition, there was a noticeable lesion on
the skin, which we classified as a typical scoop mark.
This type of skin lesion has been found in a high number of alleged abductees. There is a theory that the lesions are caused when tissue is removed at the same
time that objects are implanted in another part of the
body. We think this tissue may be used to wrap the
object before implantation, in order to prevent rejection
by the body. We were interested to see if the tough
membranes that covered the objects we removed during

the first surgeries could have been made from tissue taken in this manner.

This patient had conscious memories of an event which happened several years earlier, and said she was experiencing ongoing abduction incidents. She also had a vivid memory of seeing a UFO over her home in the San Fernando Valley, north of Los Angeles. We performed a battery of psychological and physiological laboratory examinations on her and found her to be a good surgical candidate.

The third subject, another female, appeared to offer a classical textbook case of alien abduction. She had all the signs which are described in the many books on the subject. In addition, she had a signed affidavit from her neighbor across the street attesting to the fact that on the night of her abduction, the neighbor saw a circular-shaped craft hovering over the victim's house. She was able to clearly recall many details that occurred during her abduction experience. She had a lesion on her leg which was red in color and raised. She described the history of this lesion and how it was treated by her doctor. She'd had a small surgery performed to drain the area and this had resulted in delayed healing which took the better part of a year. She said that the doctor drained a copious amount of a purple fluid and was mystified about its origin. X rays of the area showed an object similar to that of the second surgical candidate. It appeared essentially as a small opaque ball located in the soft tissues just below the skin.

I began to plan the second set of surgeries. My first job was to get estimates of the costs involved so I could put together a budget for N.I.D.S.

I went to my office one quiet afternoon when I was not seeing patients and made several phone calls. My first call was to Dr. Thomas Dodd, a friend who owned a large surgical facility in the San Fernando Valley. I told him that I was involved in a rather unusual research program and needed to rent his surgical facility for a day.

"I'm in need of some sophisticated equipment, such as a fluoroscan or a C-Arm, which I know you have," I explained.

"How many hours would you be using it and what is the date you had in mind?" Tom asked.

I gave this some thought. I did not want him to know that my plan was still in the formative stage and that I was working on a budget for a third party which would ultimately be supplying the funds.

I answered, "I was considering a date in mid-May. This will depend on the patients' availability. I expect to do three or four cases, and that requires some coordination. As far as the surgical time is concerned, I believe it shouldn't take more than one hour per case, plus time for the anesthesia and prep."

Tom said that his office manager would get back to me about the costs involved, and added, "Remember, I also have to pay the entire staff for the time you keep the suite open." I told him that I would wait for his assistant's call.

My second job was to find a chief surgeon, since Dr. A. would not be available. This wasn't easy, since I had to find someone who, while not necessarily a believer, was at least open to the idea of alien implants. I reviewed my list of possibilities and decided to call another old friend, Dr. Cass.

My friend seemed enthusiastic about helping with the surgeries but was going to be out of the country when we needed him. He suggested a young surgeon he had worked with named Dr. Pratt, who had excellent skills and would almost certainly, Dr. Cass felt, do the cases.

I immediately dialed the number he gave me. When Dr. Pratt answered, I introduced myself and told him Dr. Cass had recommended him for the surgeries. He asked me a number of medical questions pertaining to the patients and seemed satisfied with my answers. He said he would be available on the surgeries' tentative date of May 18 and that his charge would be $500 per case, or a total of $1,500 for the three cases. I told him the cost was acceptable.

A few days later I began composing a budget for N.I.D.S. The surgical costs were running higher than I had anticipated. Dr. Dodd's office manager had informed me that the cost of the operating room alone would be $1,000 per hour. The additional cost to keep the remaining part of the facility open would amount to another $300 an hour. I tried to persuade her to lower the price, without success.

Next, I called Mike Portanova, who was the official MUFON photographer and also a member of the executive board, for an estimate of costs for the still photographs and processing. He responded with a figure in the $500 range.

So far so good; my budget was taking shape. The next consideration was the psychologist's fee. I called the woman we had used for the surgeries performed in August 1995. Although she was eager to work with us again, she would no longer provide her services free of charge, which did not come as a surprise. She did surprise me by saying that a portion of the testing should be performed by a clinical psychologist. She knew of someone who could do this for us, and naturally, she told us, there would be a separate charge for her services. The total of all these arrangements would add up to about $500 per case. I was taken aback but told her I would submit her figures in my budget proposal.

Over the next few days I asked several key people to look over the budget proposal. In their opinions, all bases had been covered. I then sent the proposal to N.I.D.S.

I did not want to make further arrangements until I had some word from Bigelow. Finally, late one afternoon, while I was busy at my office, I was told that I had a phone call from N.I.D.S. It was Anne, Bigelow's secretary.

"Mr. Bigelow asked me to call you in reference to the budget proposal. The board has been looking at it and has a few questions. They would like to know the name of the patients, their medical history, and their experiences as abductees. They also would like to know if they meet your original criteria and do they have demonstrable evidence which shows on an X ray, CAT scan, or MRI?"

I was stunned at the nature and thoroughness of these questions. Once I recovered my wits I told her I would happily provide all the answers and send them by fax shortly. I typed the answers that night and faxed them to N.I.D.S.

I received another communication from Bigelow's secretary. She told me the board had approved the budget and she was going to fax me a set of documents to read and sign, if I approved of them. I received the fax, read it over, and found it a fairly straightforward agreement. I signed the documents and faxed them back immediately. It wasn't until later in the day that the realization of what had just happened began to sink in: I was going to take another surgical trip into the unknown.

I began to direct all my energies toward the project. The first step was to notify the patients that they would finally undergo the surgery. One by one, I made the phone calls. It was possible that we would perform four surgeries instead of three. The fourth potential patient was Whitley Strieber, author of the best-seller *Communion*. We had been in contact over the previous months and he'd told me he thought he had an implant in his ear. He had been traveling around on the lecture circuit and was scheduled to be in the Los Angeles area at the time of the surgeries.

I advised him that he had to have laboratory studies and an X ray of the suspected implant region done before any surgery could be performed, and said I needed this information several days prior to the event. He agreed to all my suggestions, but changed his mind at the last minute and decided not to have the surgery. He explained that he wasn't sure he wanted the implant removed, because he hadn't yet figured out what its function was and intended to have some tests done on it at a lab in Texas. I told him that I would be glad to have function studies to add to our pool of knowledge, but it was nevertheless disappointing.

I let Derrel know we would have only the three original surgical patients, all of whom had passed the battery of lab tests and physical exams that cleared them for surgery. One of our surgical candidates had a severe allergy to all local anesthetics and the only way to perform the procedure was by using hypno-anesthesia. Derrel planned to arrive in California a few days early so that he could make the proper arrangements and work with the patients prior to surgery. We decided to invite a number of people to witness the surgery firsthand, just as we had before.

I thought this was a good idea and proceeded to arrange for an additional room at the surgical facility, which had a large projection TV in it that could be used as a monitor to receive the broadcast from the surgery room. Realizing we could be deluged with people who wanted to see the surgeries performed, I decided to re-strict admittance only to those who received a special invitation. Among those invited were UFO researchers and writers, including Robert Dean, Johsen Takano (director of the Cosmo Isle UFO-Aerospace Museum in Japan), newsletter publishers Michael and Debra Linde-mann, Dr. John Mack, Budd Hopkins, Paul Davids (producer of the movie *Roswell*), Barbara Lamb, Walt Andrus, Yvonne Smith, Whitley and Anne Strieber, Melinda Leslie, and many others—almost a Who's Who of ufology.

Within a few days we started receiving responses to our invitations by mail and telephone. Many people wanted to invite friends, and I was worried that we would wind up with more witnesses than we could accommodate.

Preparations continued, and the closer we came to the actual surgical date the more nervous I got. One of my concerns was that I hadn't yet received any funds from Las Vegas, but had become personally obligated for a large sum of money. I called Bigelow's secretary and expressed this concern. She assured me that the funds would be forthcoming and that she would discuss the situation with Bigelow and get back to me shortly. As it turned out, I did not receive the funds until almost a month after the surgeries were completed. This delay in funding produced additional pressure on both Derrel and me.

The RSVPs continued to pour in. This was another of my concerns. Although the facility was adequate in size, the viewing area held only so many chairs. I didn't want a chaotic scene caused by an overflowing audience. I called Alice Leavy about it and she promised she would help coordinate the event.

The nonmedical part of the team was going to be about the same as it had been for the first set of surgeries, each individual having successfully done his or her job the time before. I thought there would have to be a few changes, however. Since all of the procedures were to be done by the general surgeon, I felt it was imperative to have a highly trained surgical nurse. I selected Mike Evans, who'd acted as our video cameraman during the previous procedures. Mike was very accommodating and agreed to sign on. I asked Jerry

Barber, who did photographic analysis for the Ventura–
Santa Barbara chapter of MUFON, to film the event. He
agreed to film the entire set of surgeries using high-tech
equipment which included specialized cameras. Some
of these cameras would be operating simultaneously
and others would be used for extreme close-up shots of
the specimens and wound areas. Another camera would
broadcast the procedures to the viewing area.

Alice arranged for another MUFON board member, Peggy Portanova, to set up the kitchen and coordinate the feeding of all our guests. Her husband, Mike,
would perform all the still photography. As with the
previous surgeries, Jack and Ruth Carlson would record
the event in writing. Bert Clemens would be responsible for the TV equipment and other electronics. Another board member, Leslie, who was on furlough from
the navy, volunteered to bring along two of her male
counterparts who would serve as security.

It was now May 12, only six days before the surgeries were to take place. I racked my brain to see if
there were any details I hadn't considered and decided
the only thing left to be done was to take the team to the
surgical site and have a rehearsal. This would familiar-
ize them with the facility and reassure me that things
would run smoothly.

May 14 was chosen for the on-site rehearsal. I con-
ducted a tour of the premises and helped everyone find
their places. There were many suggestions which we
implemented during the actual surgeries. After this dry
run, I was confident that all the bases had been covered.

On May 16, I was to pick up Derrel from the airport at 11 A.M. I set my alarm for 7:30 to give myself

extra time. That morning, the phone rang and my wife, Sharon, answered. She handed me the portable phone, saying it was a man whose voice she did not recognize.

"Hello, this is Roger. Can I help you?"

"Yes, this is Dr. Pratt. How are you? I hope this won't put you out, but I am afraid I won't be able to do the surgeries I agreed to do with you on the eighteenth. My secretary scheduled another big case and I can't be in two places at the same time," he stated matter-of-factly.

I reached forward and steadied myself against the bathroom sink. For a moment I was speechless. My mind began to race. This meant I would have to cancel the entire surgical affair. How would I explain this to the patients? What about N.I.D.S. and all the rest of the people involved? Who would cover the costs if we didn't use the facilities? I realized Dr. Pratt was still talking.

"I have a colleague who has agreed to take my place. I took the liberty of discussing the cases with him and he will be familiar with them on the day of the procedures. Please just give him a call and supply him with the address and time he is to be there. His name is Allen Mitter."

My state of shock turned immediately to calm and then to joy. My life was still on the track for which it now seemed to be destined. I got Dr. Mitter's phone number from Dr. Pratt before hanging up and called him to confirm the surgery date, then continued getting ready for my trip to the airport.

At 11, I met Derrel at the gate. On the way to my home, we discussed the particulars of the upcoming event. Since one of the patients, Doris, was allergic to all local anesthetics, Derrel would need to work with her extensively before surgery for the hypno-anesthesia to be successful.

The abduction work-ups had been completed. I told Derrel about the change in general surgeons. He took this in stride and let out a hearty laugh when I related the story. I told him I had talked with the new surgeon, who'd seemed very nice and willing to work with us. I explained he didn't want to appear on camera with a full face shot, but I didn't consider this much of a problem.

Over dinner at my house later, I gave Derrel information about the medical aspects of the surgical candidates. I brought him up to date with the results of the X rays and laboratory tests; there was nothing in the lab reports which would preclude our candidates from having the surgery performed. I told him the first case was going to be Annie. She had no allergies to anesthesia and showed a healthy physiological and psychological profile.

The second case was to be Doris. She would require Derrel's direct attention before the surgery because of her allergy to local anesthesia. Derrel told me he had made arrangements to begin her hypnotic sessions the next day, May 17. He also thought making her the second case would provide enough time for him to induce a deep hypnotic state prior to the time she entered the surgery room.

The third case would be Don, who had a small triangular foreign body in his left lower jaw. I showed

Derrel the latest X rays. Then I pulled out another X ray, this one of Don's shoulder.

"Take a look at this and tell me if you see anything strange here." He took the X ray and held it up to the light. A puzzled expression came over his face.

"Is this some sort of a joke or what?" he asked. I was grinning from ear to ear and told him no joke was intended. He pointed to an area adjacent to the left clavicle. "What in the hell is that?"

I said, "It appears to be some sort of screw. Sometimes screws are used in cases of bone fracture repair. He could have had a clavicular separation. I questioned him about this and he swears he never had a fractured shoulder or a surgery on this area."

This was just one of the mysteries we would have to contend with. Derrel asked me if we had plans to remove the screw and I told him no. I said I thought it would be a good idea to follow this patient for some time after his jaw surgery and perhaps get another X ray at a later date to see if the screw was still visible.

On the morning of May 18, 1996, we awoke early. We arrived at the medical facility at 8:05 A.M. and noticed that we seemed to be the only ones there from our group.

"Come on, Derrel, let's grab our stuff and I'll show you the layout," I said. Soon we were standing before the big wooden double doors of the surgical complex. I reached for the knob and swung the doors open wide. Derrel stepped into the waiting room and held the doors open for me as I dragged in some of our equipment. We were greeted by a friendly voice.

"Good morning, Dr. Leir. You remember me, don't

you? I'm Dr. Miller.'' I looked up into the eyes of a familiar face I had not seen for many years.

"Bill, my gosh, what are you doing here?'' I asked.

"Well, Rog, I'm working with the group that runs this place now and they asked me to be here so that I could give you any help you needed with the surgeries. I also know how to operate the Fluoroscan.''

I was glad to see him again and pleased to find there would be a separate operator for the X-ray unit. I took him aside and explained the proposed surgeries to him. I also instructed him regarding the supplies we needed and the approximate time it would take to do each case. In addition, it was necessary for me to explain about the large number of observers who were due to descend upon us soon. Just as I was making my last comments, Alice and the first contingent arrived.

Because of our rehearsal, the team already knew where to go and the individual tasks they had to perform. Soon the area was filled with activity. Cables were being connected from the surgery room to a huge large-screen TV in the waiting room. One of the cameras was mounted on its tripod and lined up with a view of the surgical table. Our professional video cameraman and his colleague had not yet arrived, and that was beginning to disturb me. Alice, Peggy, and other members of MUFON brought up huge trays of food they had prepared for lunch and placed them in the kitchen. Still other members of the team were arranging the furniture in the waiting room so that the TV screen could be viewed by the maximum number of people.

Next to arrive were our patients, who had been transported by another MUFON member of our team.

This time we made sure to pick a responsible person for the job. We greeted them and escorted them to separate rooms which would serve as waiting areas for each patient. The plan was to have our psychologist give each of them a brief examination and then have Mike Evans draw their blood to prepare the solutions which would ultimately hold the specimens.

Suddenly I heard a burst of laughter coming from the waiting area, and I knew our guests had begun to arrive. It seemed as if my presence was required in a thousand places all at the same time and I couldn't get to each one fast enough. I made my way to the waiting area and greeted the arriving guests.

Just as I turned to say hello to one of the guests, I heard a voice calling me from somewhere in the depths of the office complex. It was Dr. Miller. I followed the sound of his voice and found him in the supply room. He was inquiring about a particular item we had requested for the surgery. No sooner had I solved this problem than someone came up behind me to remind me that Jerry was not there yet with the video team. I rushed to the phone, hastily called his numbers, and left messages. It was now 9:45 A.M. and the first surgery was to start at 10. We could not begin without the correct video equipment in place.

The next person I heard calling me was Jack Carlson, who was assisting in the operation of the video feed to the waiting room. He told me there was a problem getting the signal to the TV set and he didn't know why.

"Please get it fixed," I barked at him. "There is only a certain amount of time allotted for this. We are paying a thousand dollars an hour for just this room alone.''

My mood didn't improve when I looked at my watch and found it to be 10:05 A.M. Not only was Jerry not there yet, but our general surgeon was also late. Seconds later a voice called my name from the front-office reception area. There stood a man in surgical greens. Relieved, I reached out to shake his hand.

''Allen, nice to see you. Come on in and let me show you the facility.''

Our surgeon was now present and ready to start. I took him on a tour and introduced him to the principals. I explained the videotaping to him and offered him all the anonymity he wanted. He seemed satisfied with my assurances.

By 10:30 A.M. there was still no word from Jerry. I had to make up my mind about what to do. I decided that at $1,000 per hour I was not going to wait any longer. We had one camera in operation and whatever happened, we would still have an adequate recording of the event on tape. I raced back to the doctors' lounge and quickly put on my surgical scrub suit. I spread the word to all concerned that the first surgery was about to begin. With that the back office came alive. The first patient was brought into the operating room, and immediately Mike started the surgical prep. I took a last opportunity to announce to our waiting audience that it was time to start paying attention to the TV screen.

Our first patient, Annie, was in excellent spirits, appearing calm and unconcerned about the procedure which was about to be performed on her. Dr. Mitter introduced himself and asked some routine medical history questions. I asked Mike for the local anesthetic mixture and handed the filled syringe to Dr. Mitter.

I steadied Annie's leg with both hands and Dr. Mitter plunged the needle into the spot adjoining the surgical target. Soon, the area was fully anesthetized. Allen took a needle and tapped it over the surface of the skin.

"Do you feel anything when I touch you here?" he asked.

Annie quickly responded, "No, I don't feel a thing."

I asked Mike to start handing us the sterile drapes. He reached over to the back table and handed us the first sterile covering. Dr. Mitter took one end and I held the other and we stretched it over the area above the surgical site. The second drape cover was placed below the surgical site. Next we applied a drape which had a manufactured hole in the center. The portion with the hole was placed directly over the site of the incision and sealed in place with adhesive.

We were ready to begin the surgery. The surgical Mayo tray containing all the instruments was rolled over next to the surgical site. Dr. Mitter was about to reach for the surgical blade when suddenly we were aware of activity at the entrance to the operating room. I looked up and saw a heavyset gentleman wearing a green scrub suit and surgical mask standing in the doorway with a very large video camera in one hand and a tripod in the other. It was Jerry.

"Doc, I am so sorry we're late," he said apologetically.

I was in no mood to hear his excuses, no matter

how valid they might be. I noticed he had brought his associate, Sharon. She was also dressed in surgical greens and holding another type of video camera.

"Jerry, can you just go ahead and get things set up without screwing anything up in the room? We are ready to cut the skin open at this moment. We can stop for a minute or two until you get the camera situated, but you better get it done right away," I directed.

I then reflected that my words had been carried to the audience in the waiting room, because the sound feed was live at that moment. I told Bert to shut down the live feed for a minute or two until Jerry got his equipment in place. Jerry and Sharon quickly set up the tripod and plugged in the cables and power supply units. In a matter of just a few minutes Jerry told us their equipment was operational. I instructed Bert to fire up the live feed again and apologized for the delay.

I picked up the number 10 surgical blade and handed it to Dr. Mitter. After instructing the patient to tell us if she felt anything uncomfortable, he made the first incision. I reached over with a gauze sponge and dabbed blood from the surgical site. He completed the first incision and quickly turned the blade in the opposite direction, cutting through the skin on the other side of the section we were removing. This was a routine type of elliptical incision.

We had never worked together before, and my function was that of an ordinary surgical nurse. I was glad we were off to such a fine start. It wasn't long before he lifted out the segment of flesh containing the object. I took the specimen while Dr. Mitter proceeded with closure of the wound. The close-up camera was focused on the bloody tissue resting on the sponge. I took a sharp instrument and began to probe the tissue. Suddenly I heard and felt a little clicking sound. My instrument had touched something solid. There it was: a small, round, grayish-white ball which glistened under the powerful surgical light. A feeling of excitement went through the room and I heard a cheer from the folks in the viewing room. Annie responded to our excitement by asking what we had found. I took the gauze sponge and brought it over to where she could see it.

"My heavens, did that come out of me?" she asked excitedly.

"It sure did," I answered.

Derrel stood by Annie's head and leaned over so he could get a better look at the sponge.

"That doesn't look like the others," he said. "Is that what we saw on the X ray?"

I told him it appeared similar to what we had seen on the film and that I was surprised it was such a hard substance.

While the room was cleaned and prepared for the next case, I made a brief appearance in the viewing area and answered a few questions before returning to my post for candidate number two.

I saw Derrel and Doris walking down the hall. The patient was totally unsteady and looked as if she had been on an all-night binge; I wondered if she was going to make it into the surgery room. This was the patient who had allergies to all local anesthetics and was going to have only hypno-anesthesia. I didn't know what procedure Derrel had used, but it had obviously been effective. I looked at her lying on the table and saw that her eyes were glassy. I asked Derrel if I could talk with her, and he nodded.

"Doris, how do you feel?" I asked.

"Doc, this is the greatest. I have this kind of surgery all the time and have learned to love it," she answered with a big smile on her face.

The procedures were going to be similar to the first case, but without the injection of an anesthetic. Mike performed a sterile prep as he had done on the first patient and then surrounded the area with sterile drapes. Dr. Mitter had never performed a surgery before using only hypnosis as the anesthetic. He was cautious in his approach and kept asking the patient if she felt anything when he pinched the surgical area with a forceps. She insisted she was experiencing no discomfort. Finally he seemed satisfied and gingerly made the first incision. There was a surprising lack of bleeding from the surgical site. After the surgery, Derrel explained that he could help patients control their bleeding during the hypnotic state.

The procedure progressed in a manner similar to the first case; once again, an ellipse of skin that contained the object in question was being removed. Then a terrible mistake occurred. Dr. Mitter, who had performed many surgeries under typical local anesthetic, picked up a forceps and gently grasped the segment of tissue to be cut free from the underlying anatomy. He looked directly at Doris and said, without thinking, "This may hurt just a little now, dear."

With that, the patient screamed in pain and tried to remove her leg from the surgical area. I don't know if any of us knew what had happened except Derrel, who quickly took hold of the patient's head, looked directly into her eyes, and stated in a loud voice, "Doris, listen to me. Your entire leg is frozen. It is a block of ice and you feel nothing. It is absolutely numb. It is numb and cold, just like ice."

Dr. Mitter, Mike, and I were holding on to her leg, trying to keep it in the surgical field. We could feel the tension drain from Doris's muscles, and slowly she lowered her leg back onto the table. A bead of sweat formed on Dr. Mitter's brow. Mike noticed its progress and used a sterile towel to pat the doctor's forehead dry. This was a good lesson for all of us—when a patient was having surgery under hypno-anesthesia, we would have to be very careful about what we said.

As before, the excised tissue specimen was placed on a surgical sponge. I probed it just as I had done previously. The tension in the room mounted. Again, an audible click was heard, and there it was—another small, grayish-white, shiny ball.

An announcement of our findings was made to the camera. We could hear the excitement of those in the viewing room. The wound was neatly sewn closed and a simple dry sterile dressing was applied. Derrel was still actively talking to the patient. He told her that she was going to get up and come with him into another room, where she would go into a deep sleep until he awakened her. He also suggested that she felt no discomfort and would be completely healed in a few days. She followed his commands like a robot and accompanied him to the recovery room.

The surgical crew then readied the room for the next patient. I took the opportunity to grab a quick cup of coffee with Dr. Mitter. I asked him how he thought things were going so far. He seemed satisfied with everything we had done. I noticed he kept looking at his watch and asked him if he had a pressing engagement. He told me he had promised his wife he would be home at 4 P.M. I promised he could leave promptly following the next case.

Don was on the table when we entered the operating room. Mike had performed the surgical prep and was standing ready to proceed with the surgery. The cameras were repositioned to focus on the jaw area, and Dr. Miller had positioned the Fluoroscan unit. Dr. Mitter and I were both scrubbed and ready. We donned our surgical gloves and approached the table. Mike, without being instructed, passed the sterile syringe filled with the anesthetic mixture to Dr. Mitter.

In a few moments, the patient informed us that his jaw area was numb. The drape material covered not only the patient's upper torso but also his head and face, so that when he was asked a question, his response sounded as if it came from somewhere other than his mouth.

The initial incision into the skin would be the only one made; there was nothing to remove from the superficial tissue. I helped retract the wound, holding it open so the deep tissues could be seen. The foreign body was triangular and small. It was like looking for a needle in a haystack.

Dr. Mitter looked up and said, ''I think we are ready for the Fluoroscan.'' Dr. Miller moved the arm of the unit over to encompass the jaw area. He stepped on the pedal to activate the machine. We quickly moved our eyes to the TV monitor and watched as the tissues of the jaw became visible. Suddenly, there it was.

"Hold it!" we all chimed in at once.

The small triangular metallic object was visible on the screen. I picked up a hypodermic needle and handed it to Dr. Mitter.

"Why don't we place this inside the tissue and watch it on the screen. Go ahead and try to touch the object with the end of the needle. That way we'll know precisely where to probe," I suggested.

He agreed and quickly but carefully placed the needle in the deep tissue. Slowly and cautiously, he began to push the needle deeper, a little at a time. He had to be very cautious in this area because of the possibility of hitting a facial nerve. If this happened it could result in permanent damage, including numbness or paralysis. We held our breath and watched the TV monitor. Soon the tip of the needle was at its mark; I reached over and steadied it. Dr. Mitter took the surgical blade and cut deeply into the tissues, using the needle as a guide, until he had reached the object. As usual, Mike was right there with the correct instrument, holding a clamp in his outstretched hand.

Dr. Mitter took the instrument and directed it toward the area of tissue we were viewing on the TV screen. He carefully closed the jaws and slowly withdrew the clamp from the depths of the wound. A pool of blood welled in the area, which obscured our view. I applied a surgical sponge with direct pressure to the wound, and soon it had soaked up the blood. Dr. Mitter removed the clamp and placed it directly between the jaws of the Fluoroscan. We all turned and looked at the screen. Much to our relief, contained within the tissue was the triangular metallic foreign body.

We let out a shout of excitement and instantly heard a number of jubilant voices coming from the adjoining room. I immediately took the specimen and placed it on a surgical sponge. I took a surgical blade and began to scrape the soft tissue away from the foreign body. It appeared to be triangular in shape and was definitely metal. Slowly I removed the clinging, surrounding tissue until more detail could be seen. It was then I noticed it: the metallic portion was covered with the same dark gray, well-organized membrane we had seen covering the metallic specimens we removed in August of 1995. I couldn't believe my eyes. I picked up the blade again and tried to cut through the membrane. The harder I tried, the more frustrated I became. It would not open, any more than the specimens from last year had.

Dr. Mitter finished closing the wound and applied a sterile dressing. The patient was asked if he was still hearing voices; he said they were gone. We told him we didn't know if they would come back or not but we would watch him carefully and monitor his progress. We helped him off the table and walked with him down the hall to the recovery room. He was placed on another table, where the process of monitoring his vital signs began.

Derrel and I took this opportunity to go in and greet our waiting guests. Their spirits were high and they congratulated us on our success. We asked everyone to join us in the kitchen area for lunch. Peggy had

done a wonderful job with the preparation of the food—
the display was fit for the finest of hotel banquets. The
moment she saw us coming, she came forward with a
filled plate in each hand.

We took a few bites, then set the plates aside and
returned to the laboratory. The day's work was not yet
over. There was more minor dissecting of the specimens to be performed, and we had not done the black
light examination either. Mike Portanova was on hand
to do some close-up photography. I carefully laid the
specimens on white gauze sponges. The overhead light
in the room was turned off and the room was plunged
into darkness. Derrel reached over and turned on the
ultraviolet light. I held up the first specimen and there it
was, a greenish glow in the light of the ultraviolet lamp.
The cameras were positioned close to the object and a
recording was made of the fluorescence.

We performed the same procedure on all the specimens. Each one fluoresced, but not all were the same
greenish hue—Don's was pink in color. We did not
have an explanation for this; it was just another one of
those mysteries which would have to be solved. Following the black light examination, each of the specimens
was placed in the container that held its blood transport
solution. This time all the specimens would be kept by
me. My plan was to parcel out generous portions of the
soft tissue to at least three biological laboratories. This
would mean we would have three different pathology
reports for each specimen. I would keep the metallic
and dense portions of the specimens until we knew
where their analysis would take place.

The day was almost over. Derrel and I bowed to Mr. Takano and shook hands with the remaining guests. It wasn't until some days later that we heard about the overwhelming emotional reaction that some of them had experienced in seeing the surgery.

THE IMPLANTS
C H A P T E R   N I N E

W "HAT'S in it?"
"What's it made of?"
"What's it do?"

These are three of the most commonly asked questions about the implants we removed. It should be remembered that although the questions are simple, the answers are not. Our efforts to fully answer these questions have cost us dearly in time and money. The answers have come slowly and at this time are still incomplete. We continue to wait for the results of complex testing.

The first metallurgical results came from analysis performed by Los Alamos National Laboratories. These tests, financed by N.I.D.S., were done on tiny particles taken from three of the objects removed during the first two sets of surgeries. We chose samples from the implants removed from Patricia, Peter, and Don, because these were all metallic objects that had been surrounded by thick membranes.

The first test performed was called Laser Induced Breakdown Spectroscopy. LIBS was used to determine the elemental composition of the objects because of its microsampling ability and relatively nondestructive analysis capabilities. Under microscopic examination, all but one sample were observed to have different areas that were unique in appearance. For identification, they were named the scaled, black, brown, white, and rust areas.

LIBS is an elemental analysis technique in which powerful laser pulses are focused on the tiny samples of the material under study. The microplasma vaporizes a small amount of the sample (less than 50 nanograms) and excites the resulting atoms to emit light. The light is then collected and spectrally dispersed, and the resulting spectrum is recorded to determine the elemental composition. Because each element has a unique spectral signature, the elements can be identified by analysis of the spectrum.

Different elemental compositions were found for the different colored and textured parts of the implants. Two samples were taken from Patricia's T-shaped implant and a small sliver was taken from her second, seedlike implant. All three samples were tested together for trace elements. A sliver taken from the seedlike implant removed from Paul's hand and three segments from Don's jaw implant were also tested.

The results of the testing on Patricia's objects was

as follows: the scaled portion of her T-shaped implant contained major concentrations of calcium, aluminum, and iron. The black portion contained mostly copper, calcium, and iron. The black portion of the sample taken from the same implant contained high levels of copper, calcium, and iron, just as the black portion did, although the minor concentrations of elements were slightly different. The brown portion of this sample contained major concentrations of barium and copper.

The black portion of the third sample, taken from Patricia's seedlike implant, had major concentrations of copper, calcium, and barium. The white portion of this sample revealed major concentrations of copper, calcium, aluminum, and iron. The rust-colored portion tested as being made of mainly copper, calcium, and barium.

So Patricia's T-shaped implant contained mainly iron (except for one sample), calcium (except for another sample), and copper (again, except for one sample). In other words, different colored and textured parts of the same implant contained different concentrations of elements. One sample from this implant contained aluminum as well, and another contained barium.

Her second implant was mostly made up of copper and calcium. Two areas also contained barium, and the third sample also contained aluminum and iron. The elements that made up the seed-shaped implant were very similar to those that made up the T shape.

The sample that was made up of a combination of particles taken from the other three contained the trace elements europium, ruthenium, and samarium. These elements can only be manufactured in a laboratory and are not found naturally on earth, which gave us more evidence that the objects were indeed manufactured.

Peter's object was mostly made up of copper, calcium, and iron, like two of the samples from the T shape.

Don's jaw implant was divided into three samples for testing. The first, the brown sample, had major concentrations of copper, calcium, and aluminum. The white portion of Don's implant contained the same major concentrations of elements as the brown sample: copper, calcium, and aluminum. One of Patricia's seed implant samples had this combination of elements as well, but it also contained a major concentration of iron. The third sample taken from Don's implant was described as "beads on white." It contained mostly calcium and iron.

While the number of elements that were present in major concentrations in the implants taken from Patricia, Peter, and Don were only five—copper, calcium, iron, barium, and aluminum—they were not all present in each sample. There were also other elements present in minor concentrations. These were magnesium, manganese, lead, nickel, silicon, sodium, and zinc. These different combinations of elements were quite complex. It would have been much simpler had the samples all contained the same elements.

When we received these results, we immediately contacted N.I.D.S. and asked for their recommendations. It was decided that further extensive testing should be performed. We were told their board would make a decision as to who should perform the next set of tests.

It took several weeks before we heard from N.I.D.S. again. They informed us that another world-class laboratory had been chosen for the next batch of testing. This laboratory was New Mexico Tech.

We held our breath waiting for the results from New Mexico. In a telephone conversation with John Alexander, who was directing the research at N.I.D.S., I learned there was a problem with receiving reports from the laboratory. It seems our contract with them did not include language that would allow them to give us an opinion about the test results. John told me N.I.D.S. was working on the problem and it would be resolved soon. I was beginning to understand the politics involved in scientific testing.

Finally, in September of 1996, N.I.D.S. resolved their problems with the laboratory and faxed me the letter about the results. My secretary handed it to me. I sat alone in my private office, turned on the desk lamp, and began to read. On the top, in large black letters, it said, "New Mexico Tech Letter of Opinion." I began to read the body of the report.

It contained two major statements. The first one indicated that the sample taken from Patricia's seedlike implant contained eleven different elements. The sample from the T-shaped object we removed contained an iron core. The tests also indicated that iron and phosphorus were major constituents of the material surrounding the iron core.

The second statement had to do with meteorites. They thought this was the most likely material for these fragments to have come from. On the other hand, there was a problem with the nickel-iron ratio. It seems that most meteorites contain between 6 and 10 percent nickel and none contain less than 5 percent. To resolve this discrepancy they deduced that these specimens might be fragments of meteorites.

I was astounded by this revelation and felt I had to do something to help clarify the issue. I knew that these patients had not stepped on a meteorite or hit one with the back of their hand.

I called the lab and talked with the individual who wrote the report. I told him I was going to tell him something that he might find shocking. With that warning, I said that the samples were obtained from human bodies through a surgical procedure.

As a result of this statement, the metallurgist gave us another opinion, which would ultimately prove to have little merit. The lab's final letter of opinion contained an additional general statement responding to the fact that I'd said these samples were obtained from the human body. It stated that an iron-silver mixture embedded into the body could cause a calcification reaction. It also stated that medicine and dentistry have used ceramic materials for many years. Unfortunately, the metallurgist's opinions turned out not to be true. In fact, no ceramic materials at all are used today because of the tremendous inflammatory reaction which occurs when they are instilled into the body. The author of the report did not know one important aspect of our cases: the complete lack of any inflammatory reaction. He also did not know that the specimens were covered by a strange, gray, dense biological membrane. Last of all, there was no evidence of a portal of entry for any of the implants we removed.

Metallurgical analysis has so far given us the following information: the T-shaped object is composed of two small metallic rods. The horizontal portion contains an iron core which is harder than the finest carbide steel. This rod is magnetic. The iron core is covered by a complex layer of elements which forms an outer coating. One portion of this coating has a crystalline band which circles the rod. When this rod is viewed on an electron microphotograph, it appears to be structured. One end is in the shape of a barb, while the opposite end is flat. In the center is a small depression. The shape of this depression conforms exactly to the shape of one end of the vertical rod. The vertical rod contains a similar coating but the core is composed of carbon instead of iron and is attracted to the other rod by magnetic force.

It would seem that these structured objects serve a purpose. Many people have expressed interest in studying them to try and determine their function. One person who has been working with us is Robert Beckwith, an electrical engineer. He has some theories about how these objects might work. One theory has to do with their ability to act as tracking devices or transponders. This would enable someone or something to find an implanted subject anywhere on the globe. Another possibility is that they may act as behavior-controlling devices. We know that abductees sometimes seem to have compulsive behaviors. They may wake their families in the middle of the night, ask them to get dressed and hop in the car, then drive them out into the country, where the abductee might have an encounter—or perhaps nothing happens and they just return home. I believe a more plausible purpose might be their use as a device for monitoring certain pollution levels or even genetic changes in the body. This may be similar to the way we monitor our astronauts when they are in space. Only more time, effort, and study will answer these questions.

I acquired another set of biological findings from the soft tissue specimens sent to the laboratory following the second set of surgeries on May 18, 1996.

As with the previous specimens, each laboratory was given only the bare minimum of data, so that the studies continued to be blind. Of the three objects removed that day, one was a metallic foreign body.

This was the triangular object removed from Don's jaw that was also covered with the dark gray, dense membrane. The result of the surrounding tissue analysis was about the same as with the first two surgeries, showing a lack of an inflammatory response and accumulations of nerve cells.

The results of the tests on the other two objects were somewhat different. These objects were little round balls of grayish-white material found deep within the tissues but attached to the underside of the skin. When we performed the surgery, we basically cut through the superficial skin and extended the wound deeply to encompass all the tissue below; this provided us with a good cross section of the entire segment. The results of the microscopic examination on these two objects demonstrated some of the same aspects we saw with earlier soft tissue studies, including collections of nerve tissue and the absence of inflammation.

TABLE 1. Implant Removal Summary Results

WHO	WHEN	SEX	SIDE	WHERE	COMPOSITION	FORM	COLOR	NERVES	MEMBRANE
Patricia	8-19-95	F	L	toe	metallic	seed	gray	yes	yes
Patricia	8-19-95	F	L	toe	metallic	T	gray	yes	yes
Peter	8-19-95	M	L	hand	metallic	seed	gray	yes	yes
Babs	1-2-96	F	L	shoulder	elemental solid	ball	white	yes	no
Annie	5-18-96	F	L	leg	elemental solid	ball	white	yes	no
Doris	5-18-96	F	L	leg	elemental solid	ball	white	yes	no
Don	5-18-96	M	L	jaw	metallic	triangle	gray	yes	yes
Lyla	1-12-97	F	L	heel	glass	sliver	amber	no	no
Paul	8-17-98	M	L	hand	metallic	seed	gray	n/a	yes

However, there was some new and surprising information.

One of the new findings had to do with the blood vessels in the area. Essentially the interior cavity of the blood vessels had been destroyed. The next and probably most surprising finding was that of solar elastosis. This meant that the deeper layer of the skin (the dermis) had been subjected to a large and intense amount of ultraviolet light. This seemed like an impossibility when we considered that the two female patients were housewives with no histories of extended exposure to sunlight or sunburns of the lower extremities. Even if their memories were incorrect, it would not explain why only one tiny section of their skin would show overexposure to ultraviolet radiation. There were no other marks on either leg.

Once again I took an intuitive leap toward a solution. I remembered my previous thoughts having to do with the keratin portion of the membrane. I began to wonder if an instrument existed that was spoon-shaped, sharp, and contained an ultraviolet light source. If there was such a device it would explain the presence of both the keratin and the scoop marks. It would also explain the pathological findings. Was there somewhere a being who possessed an instrument which could scoop keratin out of the superficial skin and then seal the wound with ultraviolet light? Perhaps there was secret research going on right here on earth regarding instruments capable of doing this type of procedure.

I did another surgery on January 2, 1996—between the two sets of surgeries done with Derrel—

which involved soft tissue findings that were even more difficult to assess. Babs was an alleged abductee with a painful, pinpoint-small, raised area on the back of the left neck-shoulder region. She had noticed a dime-sized, raised red lesion on her skin the morning after a possible abduction experience. She stated it was extremely painful to the touch and thought she might have been bitten by a bug or stung by a bee.

After about two weeks the lesion had shrunk to a small pinpoint area that was still painful to the touch. I considered her symptoms and in light of our previous soft tissue analysis, thought they might contradict our findings of absent inflammatory responses. However, I remembered that the tissues surrounding the implants had contained large amounts of abnormal nerve cells. This would explain why this new abductee could be having a pain response without inflammation. The initial redness could have merely been caused by an allergic response in the local tissues. This would be temporary and normally disappear within a few days.

I agreed to look into her case and interviewed her in my office, where she took the tests which Derrel and I had devised. These tests were later evaluated and scored. She scored high on the abduction probability scale. Because of these factors, I decided to take an X ray. It showed a small shadow below the area of the skin lesion.

I called a colleague of mine who was a dermatologist and pathologist. He agreed to see the patient without charge, and I arranged a time for the three of us to meet at his office. He examined the patient and came up with the diagnosis of a calcifying epithelioma. I asked him how he'd arrived at this conclusion and he explained that it was his clinical opinion based on many years of experience. I thanked him for his help and asked if he would consider removing the entire area surgically. He thought it was a good idea and suggested he do the pathological analysis himself. I said that would be fine with me but I would be responsible for the analysis of any solid object we removed. He agreed, and a date was set for the surgery.

The surgery went smoothly and we removed not only the skin lesion itself but also a little grayish-white ball attached to the superficial tissue, similar to the objects I later removed from Annie and Doris during the May 1996 surgeries. My colleague smiled and told me that the object was most probably a calcium deposit. I asked him what color calcium deposits were in the other patients he had operated on. He told me they were a bright white in color. I then asked him why he considered this object to be calcium. His answer was "What else could it be?"

I felt it was my duty to learn as much about calcifying epitheliomas as possible. I researched both dermatology and pathology texts. I found that in general these lesions usually started small, sometimes the size of a pinhead, and after some weeks or months slowly grew to the size of a dime and became painful to the touch. This was not the clinical picture presented in this case. In fact, this woman's condition was the exact opposite of what I found in the medical texts. Later analysis of the grayish-white ball would prove this lesion was not a calcifying epithelioma.

My seventh case was that of Lyla, who had an object lodged in her left heel. X rays revealed a radiodense shadow which was located approximately in her mid-heel and was superficial. The surgery was a simple procedure involving a bit of anesthetic in the superficial tissue, followed by a small incision. I probed the wound for a few minutes until I heard an audible click. This sound indicated I had touched something solid. I proceeded to expand the wound slightly, and finally discovered a shiny object. I clamped the object with an instrument and withdrew it from the wound.

After placing it on a gauze sponge, I began to examine it with a strong light and magnification. The object was narrow and elongated, measuring less than half a centimeter. It had the appearance of glass or crystal. There was no soft tissue attached to the object and no noted membrane surrounding it. I did not excise the adjacent soft tissues, as they appeared normal in color and texture.

Tests showed the object was made mainly of silicon. I decided that it was most likely an earthly object that had found its way into Lyla's heel by accident rather than abduction.

One interesting finding has been that all of these objects were removed from the left side of the body. We have investigated a significant number of other cases, and in all of them the suspect object was in the left side of the body as well.

All subjects in our study showed abnormal reactions to local anesthetics. Either the normal amount of anesthetic was not sufficient to produce the required effect or there were outright allergic reactions to these compounds.

All subjects had a compulsion to consume salty substances and found it difficult to refuse such food items as potato chips, pretzels, and other salty food. When dining, they needed to apply large amounts of common table salt.

About half of the subjects suffered from xerophthalmia (night blindness); among women, the rate was about 90 percent.

Don was the only person with an object in his body who stated prior to surgery that he heard voices. After the surgery was performed the voices stopped for a short time, but then returned. All the patients in the study appeared normal on their psychological profiles, including Don.

The preoperative laboratory tests performed on the surgical candidates showed they were all within normal limits. The two patients who underwent hypno-anesthesia showed a more rapid healing process and took less postoperative pain medication.

The nature of the psychological postoperative adjustment period varied widely among patients. One patient stated he had initially lost his psychic abilities after the surgery, but as time went on felt they were returning. One of the other patients became completely divorced from the subject of ufology and

TABLE 2. Element Concentrations Found in Removed Objects

SAMPLE	PORTION	PATIENT	Al	Ba	Ca	Cu	Eu	Fe	Mg	Mn	Na	Ni
T-1	scaled	Patricia	●		●			●				
T-1	black	Patricia			●	●		●				
T-2	black	Patricia			●	●		●				
T-2	brown	Patricia		●		●						
T-3	black	Patricia		●	●	●						
T-3	white	Patricia	●		●	●		●				
T-3	rust	Patricia		●	●	●						
T-4	black	Peter			●	●		●				
T-6	brown	Don	●		●	●						
T-6	white	Don	●		●	●						

				●		•		●	
			•						
	•				•				

T-6	beads on white	Don
RR-3	n/a	Roswell
KT	n/a	=T1,2,3

Symbol	Meaning
●	major concentration
	minor concentration
•	trace concentration
	not found

Note 1: Specimen coding differs by laboratory. Specimen KT refers to the first three T series specimens.
Note 2: Specimens RR-3 and KT were examined only for trace elements.

wanted no further involvement in it. Three of the others
continued to have additional abduction experiences following their surgeries. One of the patients feels that he
might have another implant in his body that was over-
looked.

All of the patients continue to live what appear to
be normal, well-adjusted lives. Most of them stated they
would have the surgery done again if given the opportunity.

There is still much testing to be done. If we eventually discover that the implants have isotopes not found
naturally on earth, we feel we will have proved that
some individuals with alien abduction histories have
objects in their bodies of extraterrestrial origin. From
observing the T-shaped object under the electron microscope, we now know it is clearly engineered and
manufactured with precision, rather than being a naturally occurring form. Electron microscope enhance-
ments of a silicon implant which was brought to a lab to
be tested by Whitley Strieber showed the same thing—
the implant was not a broken shard, but a cleanly manufactured object with smooth edges. We hope to soon
have enough proof to publicly present what was once
thought to be impossible: hard physical evidence of an
alien presence on Earth.

PUBLICITY
C H A P T E R   T E N

AFTER the second set of surgeries was completed, I decided to take a few days of vacation with my family. Sharon and I had some very close friends, Phil and Virginia, who lived in a small town in the Central Valley. We planned to leave as early as possible, but this particular Friday was exceptionally busy at the office and I was running late all day. We were not able to get away until about 6:30. Each time we made this journey it took the same amount of time to get there: two and a half hours.

Phil and his wife lived some distance from town, so Phil had made reservations for themselves and us at the Holiday Inn in Visalia. About twenty minutes outside of town, I called his room from the cell phone to tell them our estimated arrival time.

Our daughter, Shaina, was excited when she realized we were going to arrive shortly. She has always been a good traveler and never seems to get bored.

We pulled into the parking lot of the hotel, and soon we were knocking on Phil's door, which was right next door to ours. Both Phil and Virginia are chain smokers, and when he opened his door it looked as if we were about to enter a cloud. We took a few moments to hug and kiss before making humorous comments about the smoke. Then Phil handed us our card key and I opened the door to the adjoining room. I glanced at the telephone and noticed the message light blinking. I thought that was strange, since the only person who knew where I was staying was Janet, my office manager.

"Hey, Phil," I called out, "look at the message light on the phone. Is that one of your little jokes?"

He smiled. "Hell no, why would I do something like that? I'm hungry too and don't want to delay our dinner."

I picked up the phone and pushed the button to get my messages. In utter amazement I listened to the voice on the line telling me that I had to call back immediately, as they wanted to do an interview with me in regard to the implant surgeries. I related this message to my wife and friends, and asked them what they thought I should do. They agreed that I should call back and see what this was all about. I dialed the number I had written down and when a female voice answered, I identified myself.

"Hi, Doc," she said. "I'm Vicki from *UFO Magazine*. We heard all about the surgeries you performed and I wanted to do an interview with you. We have a deadline, so I would like to get it done right away."

Astounded, I replied, "Vicki, I'm flattered you want to write an article about this. How soon do you want to do the interview?"

"Can we do it right now? It won't take very long."

I asked her to hold on for a moment and then told the group what was going on. I suggested they go on ahead and I would meet them in the restaurant.

I was so hungry I was ready to eat the pillow off the bed, so I asked Vicki to make the interview as short as possible. As it turned out, we were on the phone for more than an hour, with questions left over to be answered at a later date. I hung up and looked at my watch. I knew I was in trouble with the rest of the group and hurried downstairs to meet them.

At dinner, Phil and Virginia asked if I had done many interviews about my work with abduction victims. I said this had been the first formal interview and that since it was for a UFO publication, I wasn't surprised that they'd heard about the surgeries. By the time we left the restaurant and headed for our rooms, it was almost 2 A.M.

I dreamed a telephone was ringing. With each ring I became more awake, until I realized the sound I was hearing really was the phone. I reached over to the nightstand and groped for the receiver. I couldn't imagine who would be calling. I glanced at my watch—it was almost 9 A.M.

"Hello," I mumbled into the mouthpiece.

"Hey, buddy, aren't you up yet?" I heard Phil say, as if he had been awake for hours.

"What's on your mind so early in the day?"

"I just thought you'd like to know that some guy is looking for you. I guess you're famous now."

"Is this one of your jokes?"

He started laughing and I thought for sure he'd found some excuse to wake me up for no good reason. Suddenly he became quite serious and told me a man with a radio broadcast service had called and wanted to talk with me about the abduction phenomenon. He was put through to Phil's room by mistake, since the reservations for both rooms were in Phil's name. He told me he had taken the man's number and I was to call him back around 2 that afternoon.

I was still not quite awake and needed some time to digest what I had just heard. Sharon looked at me with a questioning expression on her face. Shaina had heard all the commotion and was also awake. She reached for the TV remote and tuned in to a children's program. I told Sharon what had just taken place. She felt the entire episode had a dreamlike quality.

I didn't believe that anyone who was not intimately involved with the UFO field would be calling me about the surgeries. When I called the number I had been given, I was in for a surprise. A man came on the line and asked if I was the doctor who performed surgeries on patients to remove alien implants. I didn't know how to answer him, since I was still trying to conceal my identity from the public. I asked him how he found me, but couldn't get a straight answer out of him.

I admitted I was the person he was looking for, but said I wanted to keep my identity concealed for professional reasons. He told me I was hot news. He worked for a radio news hot-line service, and with my permission would get me on radio stations all over the world to tell my story. If I wanted to remain anonymous, I could use any name I chose. I told him I would consider his offer but would have to consult my partner first. He seemed satisfied and told me to contact him again within twenty-four hours.

I told my wife and friends what was going on and they encouraged me to go through with it. Soon I was on the phone with Derrel, explaining about my contact with the radio people. He told me he had been on many radio shows, including *Alt Bell's Coast to Coast,* and that I should go ahead with the interview. I told him I still wanted to keep my true identity secret.

I realized I needed to call Vicki at *UFO* and tell her to give me an alias. I didn't realize at the time that this was only the beginning of a long chain of events which would ultimately result in my name being revealed on national television.

My next phone call was to the radio service. I advised them of my decision to go ahead with their plan but said I wanted to be known as Dr. X. They agreed and explained that within a short time I would start to get calls from various stations around the world. The rest of our trip was quite ordinary. We enjoyed our visit and on Monday morning said our good-byes and headed for home.

The very first thing Sharon would always do when we returned home was check our answering machine. Once all the baggage was inside and my wife finished with the phone, I asked her if there were any messages for me.

A pencil still in her hand, she looked at me and said, "I think you better have a look at this. There is an entire page of phone calls for you to return, including two calls each from KNBC-TV and FOX-TV. I don't understand why they're calling you."

I was in a state of disbelief. I still did not understand the importance of what I had done, but evidently someone out there did.

I took the list from her and began to read the messages. Besides the calls Sharon had referred to, there were messages from both local and out-of-state radio stations. I decided I would wait until the next day before getting in touch with anyone.

In the months that followed, I found I had so much to learn that I began to feel like I was back in school again. By January 1996, I had been interviewed on more than fifty radio programs worldwide. I gave up on the idea of an alias, and went ahead and used my own name. I was just starting to get the hang of it. Some of the shows were obvious setups, with me as the dupe and the host out to get the most entertainment possible. But I was a fast learner, and I began to be able to get a feel for what was about to happen on any given program.

Sometimes a talk show host began to make comments such as, "Well, Doctor, so you went poking around in a toe for these alien implants? Did you expect you were going to be taken up in a spaceship?" This kind of question told me this was a host who needed to be carefully controlled or I would find myself in an embarrassing position.

I would usually come back with something like, "Let me ask you this: When you were on your way to work this morning to do this show with me, did *you* think you might be taken up in a spaceship?''

This type of comment would usually stop the host for long enough to shift some of the control back to me. At that point, I wouldn't wait for his answer, but would continue on in a very scientific manner, which usually kept him quiet until I was ready to stop talking. I've found that radio show hosts will let guests continue to talk endlessly unless there is a commercial break coming up. They need to fill up their airtime, and ''dead air'' is a bigger problem than what a guest may or may not talk about.

I soon became familiar with print and television as well. I had almost stopped my writing for the MUFON newsletter altogether; there were just not enough hours in the day. I continued my practice full-time and tried to somehow fit it all in. There was a continuous stream of reporters calling me, wanting to do a story for this paper or that magazine. Some of these were domestic and some foreign, and I never knew who was going to call next.

One night I was attending a meeting of the Los Angeles chapter of MUFON. During the proceedings my beeper went off. I left the room to find out who needed me. It turned out to be a radio station in South Africa. They wanted to do a show via telephone in half an hour and asked me for a number where they could call me. I told them there was only a pay phone where I was and gave them the phone number. At exactly 9:30 P.M. I was on the air via telephone from Burbank, California, broadcasting to a South African audience.

About a month later I received a call from Walter

Andrus, the international director of MUFON. I had talked with Walt several times over the years and had developed a good relationship with him. He asked me to do a presentation at MUFON's annual symposium. I said I considered it an honor to be asked, and told him I'd gladly appear.

The conference was to take place in Greensboro, North Carolina, from July 5 to 7, 1996. I had to write out my talk so it could be published in the symposium proceedings. I knew this was going to occupy a major part of my time. Being invited to the MUFON symposium seemed to trigger other invitations, and they started to come in like a high tide. I was invited to speak at conferences in Florida, Illinois, Texas, California, Oregon, and Washington. The more conferences I went to, the more I was invited to.

Soon television shows started calling. I appeared on many TV programs, including *Paranormal Borderline, Strange Universe, Fox News, CBS News*, and a six-part series on Iranian television.

I began to receive phone calls from programs such as *Hard Copy, Sightings, 48 Hours,* and dozens of other shows. I learned to be careful not to make any kind of a deal with show representatives. They all wanted to do the so-called balanced piece, which meant they could bring on skeptics and debunkers. I wanted to make sure any information was presented in a clear-cut scientific manner.

In July 1996, I left for the MUFON symposium in Greensboro. I expected the weather to be hot and humid and was pleasantly surprised to find the temperature quite mild. The hotel was full, and I recognized many familiar faces. At this conference, I learned what it felt like to be in the inner circle of the UFO field. I also learned a few other things, such as the meaning of words like "rivalry," "greed," and "jealousy."

My presentation was well received. There was a question-and-answer session, and some of the questions asked were less than friendly. Some pertained to Derrel. He had been an active researcher for twenty-seven years and during this time had made many friends, but had also acquired some enemies. It came as quite a shock to witness the nasty behavior that existed in such an important research field.

My education continued. The more I became a public figure, the more criticism I received. I learned there were a certain number of armchair experts in every field of science or pseudoscience. Every one of them had access to the Internet. They used this medium to gossip, dissect, and make comments on other people's research efforts. Some of these so-called experts had a little knowledge in one field and some had knowledge in several fields, but they *all* seemed to be experts in our field of ufology.

There were also moments of sheer joy. One example of this occurred when I arrived in Tampa for a conference. I had just left my flight and entered the terminal. As I walked through the gate, I heard my name being called.

"Dr. Leir?"

I saw a small crowd of people up ahead. A short, pretty lady in a green dress stood there waving frantically. I didn't know her. As I approached, she darted forward and said, "Hi, my name is Anne. I was on the flight with you and thought I recognized you. I am so happy to meet you. Would you please give me your autograph? I'd also like to introduce you to my friends. They will be attending the conference and would be thrilled if you would give them your autograph.''

It is difficult to describe my feelings. I couldn't believe this was happening to me. How in the world had I gotten to the point where someone would want my autograph?

The year 1997 was spent traveling to many countries presenting our work and our findings. I have come to the conclusion that the alien abduction phenomenon is happening worldwide. During our trips, we were given many different specimens from either UFO crash sites or implants from the bodies of abductees. The results of our testing have come in a little at a time, slowly providing new clues to the puzzle.

Derrel and I were invited to present our work at the opening of the new Cosmo Isle UFO-Aerospace Museum in Hakui, Japan. It was a great honor. We were two of seven researchers invited to represent the United States; the others were Dr. Bruce Maccabee, Dr. Jesse Marcel Jr., Dr. Richard Haines, Dr. Leo Sprinkle, and Colin Andrews. We found the Japanese people to be very ceremonial, polite, and extremely respectful of the work we were all doing in the field of ufology. The museum itself is a brand-new building devoted entirely to the subject of UFOs. There's nothing like it in the entire world, and I believe it will become a world center for UFO research.

One of the most phenomenal countries we visited was Brazil. The people there were wonderful. Anyone who is not convinced that abductions are real should visit this country. Derrel and I stood outside the auditorium where the conference was held from about 9 A.M. until 9 P.M. talking with alleged abductees. Often with the help of an interpreter, they all told us their stories.

Most of these stories were similar to those we'd heard in the United States. The beings they described were the same. However, the abduction phenomenon was different for some of the Brazilian Indians who lived in the rural jungle. We were told about cases there which could best be described as human mutilations. Some of these reminded me of cattle mutilations.

Could anyone who has witnessed the things I have seen and talked to the people I have talked to doubt that this phenomenon is real? I believe that if scientists could see what I saw in Brazil, their reservations about ufology would end.

Weeks later, we were back at N.I.D.S. headquarters. The sun was beginning to set, and cool, crisp breezes swirled around us. As we walked along, the smell of honeysuckle permeated the air. Although the day had been long and tiring, I was left with a feeling of happiness.

Derrel and I spent most of the afternoon in the dining room of Bigelow's splendid building, waiting for a response from the board of N.I.D.S. with regard to specimens we had just presented.

We had traveled to Las Vegas with the implants and tissue samples that had been removed from Annie, Doris, and Don during the second set of surgeries. Now, while we sat and waited, we received many short visits from Bigelow, as he explained the progress the board was making toward acceptance of our new material for analysis.

It was not until the very last meeting with him that we realized we had an agreement. I had decided beforehand that we needed any agreement, as well as any release for the specimens we were about to turn over to him, to be made in writing, but during our final get-together, it became clear that this wasn't going to happen. Bob was so congenial and seemed so sincere that he convinced us a handshake agreement was enough. At that point we did not know if we were turning over anything of monetary or scientific value. Also, we were in no position to refuse his offer if we wanted to continue our work.

Afterward, I again found myself strolling through the magnificent gardens behind Bigelow's office complex. The pace of both our walk and our conversation was set by Bigelow. He would ask a question or make a statement and then he, Derrel, and I would stop dead in our tracks to consider either his statement or our answer. We also took this opportunity to ask him questions.

"Bob, how are we going to handle the mechanics of getting the actual funds for our continued investigations and surgeries?" I asked.

He began strolling again. Suddenly he stopped, turned in my direction, and said, "What will happen is that N.I.D.S. will issue a check payable to you and then you'll deposit it into an account. By doing it that way you will be able to keep track of the money you spend. Of course, you will be responsible for any tax liability."

I listened carefully to what he said and replied, "Bob, my suggestion would be to make the check payable to our nonprofit corporation, the Fund for Interactive Research in Space Technology, or F.I.R.S.T. This way we can keep better track of the funds and we won't have to pay unnecessary taxes."

He seemed satisfied with that arrangement. We resumed our pace again, and when we had almost arrived back at the main building, Bigelow stopped, turned to us, and said, "I want both of you to understand we have many things to do before we actually send the specimens out for the first set of tests, and this will take some time. Please be patient. We will stay in touch with you."

After our visit to N.I.D.S., the main thing Derrel and I had to talk about was F.I.R.S.T., which we hadn't actually set up yet. We had been slowed down by our financial situation. Some of the documentation was in place but we still needed money to pay the attorneys. We agreed this should become a priority issue.

The thought of forming a nonprofit organization had occurred to us within a few months after the first set of surgeries. We'd quickly learned that appropriate tests had to be performed by qualified persons who were able to generate reports on the letterhead of the institutions where the tests were being performed. The only way to get this type of testing done was to pay for it. We needed a way to raise funds for our research. We could not continue to pay for the surgeries and everything else that went along with them. We decided that the solution was to form a nonprofit organization which could raise the necessary funds.

The goals of our organization would be mainly to raise money and to inform the public about our work. We also wanted to publish our test results in scientific publications. We realized it was going to take some money up front to get the necessary documentation and tax status filed. Among my contacts were two barter exchange companies in which I participated through my podiatry practice. I decided to give them a call and see if they had someone who did this type of work.

The next morning I made a series of telephone calls to both the barter companies and the attorneys they recommended. One individual sounded just right for the job. He was located in the city of Ventura, which was only about thirty minutes from my office. I called and made an appointment for the following day.

Ventura is very close to the ocean, and that morning the sky was obscured by a thick layer of clouds. When I parked and stepped outside, the smell of the damp, salty ocean air filled my nostrils.

I entered the restored 1920s-style building and took the old wooden elevator to the third floor. The hallway was dimly lit and there was a smell of age that added to the nostalgic ambiance of the building. I found an office door with "Howard Sale, Attorney at Law" lettered on the stippled glass. As I pushed the door open, I heard the sound of a bell. The room was decorated in a type of period furniture in keeping with the age of the building. I thought for a moment I had stepped through a time portal and fully expected to find an old-fashioned female receptionist sitting behind a desk to greet me. Instead there was absolutely no one. I wandered about the room, peering at the old books in the wooden book-cases.

Soon I heard footsteps approaching and saw an older, gray-haired gentleman clutching an unlit pipe. "Hello there, I'm Mr. Sale. You must be Dr. Leir." He gestured for me to follow him into the inner office. The decor was similar to the waiting room, filled with antiques. He settled back in his oversized chair and lit his pipe.

"Let's take a look at the papers you've got there."

I reached into my briefcase for the documents I had filled out. He leaned back and started thumbing through them. It took only a few moments before he looked up at me again.

"So you want to form this nonprofit corporation? It all looks pretty aboveboard to me. I don't see any problem."

I took that as good news, but thought I had better get some idea of the costs involved.

He grinned widely and said in a low voice, "You know, you've come to me through the barter system, haven't you?" I nodded, and he said, "My fee will be on the barter, but you guys have to come up with the filing costs and the initial tax amount. The filing fees are nominal, but the initial tax deposit in this state may run you as high as a thousand dollars."

This was a shock. We didn't have that amount of money, and I began to think that we were going to have to pursue other avenues. Then he took a long drag on his pipe, reached over, and opened a very thick book.

"There is another way we can do this, and that's by registering your corporation in a state that requires less money. After it's registered there we can come back to California and register it as a foreign corporation.'' That sounded fine to me.

Within the next two months, we had our corporate seal and all the other pieces of legal and tax paperwork had been completed. Our nonprofit status was approved, and we opened an account at a local bank. F.I.R.S.T. was born; we were ready for Bigelow.

SECRETS
C H A P T E R   E L E V E N

WEBSTER'S Dictionary defines the word secret as "something kept in concealment; anything unrevealed or kept from the knowledge of others; a hidden cause; a mystery."

When it comes to scientific research and UFOs, I don't believe in secrecy. Despite this, I have at times been forced to keep some of my work secret, at least for a little while. My goals have been thwarted by the necessity to get funding. This has put me in a position where I've had to comply with the will of others. I had originally decided that all information I collected would ultimately be communicated directly to the world. So far I have been able to keep that commitment. However, it has taken much longer to acquire this information than I thought it would, primarily due to the agreements I was compelled to make along the way.

During our initial unsatisfactory experiences with the free research we had performed for us, we realized that raising money would become vitally important if we were going to be able to get adequate testing done. We discovered that money usually came with strings attached to it—when individuals gave us money, they also made demands. We learned that there were sacrifices we would have to make in order to achieve our goals.

Our first exposure to secret agreements came about at our meeting with Robert Bigelow. It didn't take him long to make it clear that we would have to fulfill certain requirements before N.I.D.S. would agree to any type of funding. One of the most important of these was not to divulge who was working on these implant specimens. In the beginning, he emphasized that not even N.I.D.S. was to be mentioned. In addition, he requested that the names of the laboratories they were using for the tests also be kept confidential. This turned out to be one of the easier promises to keep, since N.I.D.S. did not tell us who they sent the samples to. The third part of our arrangement was a request for us not to release or publish any analysis results until we received clearance from N.I.D.S. At that point they would also advise us about where they thought we should publish.

These points were difficult to agree to, but I carefully weighed them against the benefits Bigelow offered, which were an unlimited supply of funds, use of the finest laboratory facilities, and peer review by the

N.I.D.S. board, which included some of the best scientific minds in the United States.

We came to the conclusion that we had no choice but to carry on with the proposed plan. It wasn't until we were well into the investigative process that we really began to feel the effects, however. I was the first victim.

As I mentioned earlier, I was asked by the international director of MUFON, Walter Andrus, to do a presentation at the National MUFON Symposium in July 1996. As I proceeded to plan my lecture, I became aware that the best I could do would be to tell only part of the story; I would have to keep N.I.D.S.'s participation in the project a secret. This was frustrating, and didn't jibe with my personal policy of releasing all my scientific data to the public.

I also thought that Derrel should be at the symposium with me, since it was his story as well. I immediately called Andrus to give him the rundown on my progress and to ask him if Derrel could also be present.

"Well, Roger, you know our organization has only so much to spend, and I don't think we can afford to pay the transportation and accommodations for another speaker," he responded.

I tried every argument I could think of to get him to change his mind, but nothing worked. Then came my final suggestion.

"Walt, I think if you could just pay the airfare, we could stay in the same room and would share the honorarium between us."

When he didn't respond to my suggestion, I sensed there was something he wasn't telling me. "Walt, what

is the real reason you won't invite Derrel?" I asked
bluntly.

He reluctantly admitted, "To be frank, I think there
is some question about Derrel's reputation."

I was shocked. So far as I knew no one had ever
said a bad word about Derrel. I asked Walt to explain,
and he alluded to an article that had been written some
months ago in the *Houston Post*.

I immediately called Derrel for an explanation.
When he answered the phone I didn't waste any time
before asking him about what Andrus had told me. He
listened and then supplied me with the story behind the
newspaper article. Apparently Derrel had taken a stand
against a Houston-based UFO organization that required alleged abductees to have a psychiatric examination in order for them to become involved with the
group. He angered some people within the organization, who then set out to discredit him. This was when I
began to learn how much gossip there is in the UFO
community and how harmful it can be.

I went to bed that night with the day's events
weighing heavily on my mind. All I could think about
was the word "secrets." I had the sense that this was
going to be a strange night.

Suddenly I went into a deep, trancelike sleep. Without warning I found myself standing in what appeared
to be a courtroom. There were wooden chairs to my
right and left. Lying on the table was a withered brown
book covered with a thick layer of silvery dust. I
reached down and touched the decomposing cover. It
felt as old as it looked. As I withdrew my hand I noticed some light silvery powder clinging to my fingers.

Suddenly I heard a noise behind me. I was terrified and confused. Finally I got up the courage to turn my head and look behind me. I saw about a dozen shadowy figures slowly entering the room. One by one they began to take seats in what appeared to be a gallery. They were all about the same size, under four feet tall, and were wearing shiny, silklike garments which seemed to wrap around their bodies. They had sashes belted about their waists and hoods thrown over their heads and draped over their shoulders. As they slowly shuffled to their seats, the room fell into a complete and deafening silence.

I turned back to the front and was startled to see a figure standing there. All at once the sound of a deep, booming voice resonated through my entire being. The same words repeated over and over again.

"Secrets . . . secrets . . . secrets . . . look, listen, and learn."

The repetition was like the beating of a bass drum. I asked myself, "Is there a lesson here for me?"

The figure in front of me became easier to see. He was taller than the others and wore a similar garment, although the material did not appear to be as shiny.

I tried to compose myself and use my scientific mind. I attempted to use all of my senses to observe. At that point I noticed ancient, withered white fingers clutching the front of the podium behind which the creature stood. Suddenly he said, "So, you want to know the secrets. Why do you persist in this?"

I couldn't believe my ears. Was he addressing me?

"Hey, wait just a damn minute, I didn't ask to be here and who in the hell are you, anyway?" I bellowed.

To my surprise, I heard a chorus of voices. I strained to listen to what they were saying. I was able to make out a few distinct words: "science," "learn," "seek," "care," "love," "watch," "listen."

Within moments everything began to blend together in a whirring, blurring hum. The pitch was rising higher and higher. My head began to swim. I felt dizzy and slightly nauseated.

Then I realized I was back in my room, in bed and lying next to my wife. My daughter had climbed into bed with us and was sitting straight up, staring at me. I must have had a surprised expression on my face, because she leaned over and whispered in my ear, "Daddy, where were you?"

What in the world did she mean? I looked at the clock. It was almost 4 A.M. I said, "Lie back down and go to sleep, sweetheart—it's late. We'll talk in the morning."

I thought deeply about my nighttime adventure and wondered if it was a warning. At the least, it changed the way I felt about keeping secrets. Maybe the lesson I needed to learn was that secrets are part of our very existence and will always be with us. Essential human questions, such as the truth about God and life after death, are still cloaked in secrets, and perhaps always will be. There is no harm in searching for the truth, however—something every good scientist struggles to do.

My next exposure to the realm of secrets came some months later. I was given a piece of material that was supposedly taken from the Roswell crash site. (My reputation for using scientific methods was becoming well known. Many people who had materials acquired from a variety of sources having to do with the UFO phenomenon were beginning to contact me. Some of these individuals feared for their lives.) I agreed to make no disclosure about who had supplied the Roswell material and agreed to report the results only to the owner.

I proceeded with the analysis and at a certain point decided the results should be made public. I contacted the owner of the object and asked his permission to divulge the composition of the sample. He surprised Derrel and me by giving his approval.

I decided to release this information in July 1997. I got in touch with our friend Chris Wyatt, who had formerly worked for CBS, and he decided to hold a press conference in Roswell, New Mexico. There would be a huge celebration of the fiftieth anniversary of the Roswell UFO crash on July Fourth, and this would attract news media from all over the globe. I was convinced that we would be making a statement that would shock the world.

One of the scientists involved with the analysis was Dr. Russell VernonClark of the University of California, San Diego. It took a great deal of courage for him to come forward and put his reputation on the line. The purpose of the press conference was to make the incredible announcement that we had a piece of material believed to be from the Roswell crash site, that it had undergone scientific analysis, and that 99.9 percent of it was found to be silicon containing extraterrestrial isotopes—in other words, that it was material that was not made on this planet.

Our press conference took place on schedule, but the results were less than satisfying. This was the same day that the Mars Pathfinder was scheduled to land. Every TV and radio station was covering that extraordinary event, so we didn't get nearly the attention we'd hoped for. We also learned that the AP news service had been less than kind to us. They did make an effort to verify the employment of Dr. VernonClark at the University of California, but in doing so they supplied the university with the wrong name and consequently were told that no such individual worked there. As soon as we became aware of the situation, we contacted the wire service and told them about their mistake. They apologized and said they would print a retraction. Technically, they did as they promised. Their original statement appeared in over sixty newspapers, however, and their retraction in only two. This gave me instant insight into the workings of the print media. I doubted that the whole affair had been just a simple mistake.

This was another one of the lessons I had to learn. Not only were we working in a field which had never been seriously scrutinized by the scientific establishment, but we were also at the mercy of the media's unwritten instructions to debunk UFOs.

Since then, I have come across information exchanges among scientists who were debating portions of the Roswell data presented by Dr. VernonClark in July of 1997. Several of the scientists who looked at it had questions about the mathematics involved in the silicon isotopic ratios. As it turns out, one of the people who made the calculations had erred. The net result was a new calculation that showed an even greater deviant from earthly isotopes than previously recorded.

In early 1997, Derrel and I were in England to present our material in the cities of Andover and Bournemouth. We were guests of a noted rock-and-roll star, Reg Presley. He had been a member of the Troggs, who were extremely popular in Europe during the 1960s. Presley was extremely interested in the field of ufology and was generous in his financial support of UFO events.

During our stay in England, we were treated to interesting tours of the English countryside, as well as a memorable trip to London. We were introduced to Ray Santilli, owner of the famous *Alien Autopsy* film that has been shown on Fox television.

Presley had arranged a meeting with Santilli at a Chinese restaurant. During lunch, Derrel and I offered to help Santilli authenticate the mysterious autopsy film. Our proposition was simple: If he would give us a few frames of the original film, we would have the chemistry analyzed at a credible laboratory. By using this method we would establish, once and for all, an accurate date for the film's manufacture. We knew that Santilli had been criticized for the way he had presented the film without objective analysis. The results of the testing would initially be sent to us at F.I.R.S.T., then we would forward them to him and he could handle the data in any way he saw fit. It would become Santilli's decision to either keep the results of the analysis private or release them. Santilli didn't take us up on our offer.

Some months later, Derrel was back in the U.K. for another conference. During this visit, he was given a package by Presley which contained pieces of film from Santilli. The package also contained a letter from Santilli explaining the nature of the film scraps.

After Derrel arrived home he told me about the film. I was surprised, because I hadn't expected Santilli to accept our offer. We searched for the proper place to have the tests performed. We turned over the samples to the lab, and while the results of those tests still fall under the category of "secrets," I *will* say that they did not convince me that the autopsy film is genuine.

It came as a big surprise when I was asked during a radio interview about the autopsy film samples. This was supposed to be confidential information. How could anyone have known?

Soon after the radio program aired, the Internet was filled with comments and criticisms of the entire event. During the broadcast I'd mentioned our luncheon with Santilli, and explained that we had celebrated his birthday. I was shocked when I received an E-mail from one of Santilli's friends in England saying that my entire story was false because Santilli's birthday did not occur during that time.

I asked the English ufologist Philip Mantle for his help in the matter, as he knew everyone involved. Since Reg Presley was present at the luncheon, Philip was going to call him and ask him about it. Reg said he had no memory of celebrating Santilli's birthday. When Mantle E-mailed me with this information, I began to doubt my sanity.

I told Philip again about my memories of details that included a champagne toast and singing "Happy Birthday." Perhaps it was my insistence that made him call Presley again. Presley was on his speakerphone when Philip asked him the question again.

"Reg, are you absolutely certain that during your luncheon with Santilli you never celebrated his birthday? Roger and Derrel both insist this happened and describe your singing 'Happy Birthday' and making toasts with champagne."

Presley answered, "Philip, I am sorry, but I don't recall any birthday celebration for Ray."

Philip went on to tell me that he suddenly heard a burst of laughter, then Reg excused himself for a moment. Philip said he could hear a female voice shouting in the background, "Reg, what's the matter with you? That was *your* birthday!"

The voice was Presley's wife. She reminded him that he had celebrated his own birthday on Sunday and told him he must have also celebrated it at the Santilli luncheon on the Saturday before.

The matter was instantly cleared up and our sanity was restored. I wasted no time in sending an E-mail to the writer of the critical letter, and in a very short time I received an apology.

I considered this another lesson in the subject of secrets. I have seen misunderstandings or even outright lies get started on the Internet and then get passed around at an astounding rate. As trivial as the birthday confusion was, it's a powerful example of what can happen when secret information is leaked to the general public, as UFO information so often is, instead of being presented in a forthright, scientific manner. Distortion almost always occurs, causing *all* of the information— both the true and the false—to be debunked.

TELLING THE WORLD
CHAPTER TWELVE

I
T didn't take us long to learn that it was one thing to gain knowledge but another matter altogether to make that knowledge public.

Derrel and I had promised ourselves that we would present whatever knowledge we gained from our research to the entire world. It seemed like so many in the field of ufology had done fine research, but when we read through the literature, we found that a great deal of the data seemed to be unavailable to the general public or, in some instances, even to other researchers.

We were unable to determine whether this occurred due to a lack of interest on the part of science and the media or if the information was being kept secret on purpose. One example of this is the abduction literature itself. I was amazed to find how little detail was available having to do with the driving mechanism of UFO craft. Where was all the information on the temperature of the floors and walls that abductees came into contact with during their experiences? Were Betty and Barney Hill the only ones who were able to identify a star map? It seemed to me that a lot more individuals should have seen these kinds of clues.

When it comes to the description of alien captors, it is an entirely different matter. Everyone seemed to be writing about the Grays, the Nordics, the Reptilians, etc. Perhaps this is because these are the kinds of descriptions that excite the tabloids.

I decided to contact some researchers in person. When asked point-blank about these details, some answered by telling me there was a certain amount of information they had to restrict from the public. This was necessary, in their opinion, because it helped to determine who was telling the truth about their abductions.

We found another reason for restricting information, having to do with personal and petty reasons. In order to get their material out first, researchers sometimes don't tell everything they discover so they can profit from their discoveries. If scientists treated their new information this way, little progress would be made. When I learned how common this practice was in ufology, it upset me deeply. I could not understand the mentality of individuals who withheld information for those purposes. Surely if their discoveries were validated by other researchers, this would lead to the kind of verification which would help the public to take the

UFO message more seriously, and this would benefit everyone in the field.

We got so many surprises from our test results that I began to doubt my own knowledge. In order to assure that our investigation was heading in the right direction, I had to be certain the information we were receiving was accurate before we issued statements to the public or scientific community. I spent many evenings sitting at my computer delving into the texts of on-line medical libraries. This was followed by discussions with my medical colleagues.

As we began to release our findings, we were besieged by lecture promoters, magazines in the UFO field, radio programs, newspaper reporters, and, finally, television offers. At the same time, some very reputable scientists approached us with offers to perform research if they would be allowed to share the authorship of an article in a scientific journal. We readily agreed, as this was part of our original plan.

Within a few months, I had spoken on more than one hundred radio programs. Some of these were with Derrel, others I did by myself. Robert Bigelow had his own ideas about releasing our information. When we originally entered into the agreement with him, he'd stipulated that articles had to be published only in the nation's leading scientific journals. Therefore I was surprised when he called to ask my permission to release some of the scientific data to the *MUFON Journal*. Apparently, Walt Andrus had asked him if he would be willing to release the information. My positive feelings about MUFON were well known, so he probably was surprised when I told him not to proceed.

The reasons for this decision were many. First and foremost was the fact that the research was not complete and I did not want to be in a position of being forced into premature conclusions. Another reason was that I wanted to personally write the article, so I could be sure that it would not only present the findings correctly, but also give the background on the research itself. I was finally ready to write the article in the spring of 1996, and it appeared in the April issue.

After completion of the second set of surgeries, we had a large amount of videotape available. F.I.R.S.T., our nonprofit organization, was still in need of funds to perform further surgeries and continue research on the metallic portions of the samples. It occurred to us that we could possibly use the entertainment industry, particularly television and motion pictures, to get funding for our projects.

At the time it seemed simple. All we had to do was find a producer who would create a work about what we were doing and use some of our footage. So we set out to find the right individual. Our travels took us to Hollywood, where we had meeting after meeting. Some were a total waste of time, but others eventually had a degree of success.

We realized we needed professional help if we were going to be able to move our project forward. I've lived in southern California since 1948, and at one time I was even employed by the motion picture industry. I discovered that the way it does business has not changed since then. Deals were then and are still made in studio offices, restaurants, hotel lavatories, parking lots, and backseats of cars. The common wisdom was that you could not trust anyone in the entertainment industry and if you did not get your money up front, you would never see a dime.

William Morris, one of the most prestigious talent agencies, became our next target. I made a few telephone calls and cashed in on some favors I was owed, and eventually got the name of someone I could talk to within the agency. The more famous Hollywood agencies are not places where you can simply walk in off the street and get an interview. It all had to do with whom you knew and who might owe you a favor.

I made the call and asked for the secretary of Steven Farnsworth.

"Mr. Farnsworth's office, may I help you?" a charming female voice asked.

"Yes, my name is Dr. Leir and I am a friend of Maxine Klein. Would it be possible for me to speak with Mr. Farnsworth?"

I patiently waited until she came back on the line. "Dr. Leir, Mr. Farnsworth is in a meeting. Can he call you back this afternoon?"

"Yes, that will be just fine. Let me give you a number where I can be reached after two o'clock. Please make sure he calls me after two."

"I will put this on his desk and have him call after two o'clock as you requested."

I hung up the receiver and wondered if I would ever hear from him. Being "in a meeting" is a common way for an agent to let you know he doesn't want to talk to you.

As it turned out, I did receive a call that afternoon from Mr. Farnsworth. He appeared to be interested in what I was telling him and indicated he would take the information to his superior, then get back in touch with me.

I was surprised again a few days later when I received a call from the William Morris Agency with Mr. Farnsworth on the line. This time he was even more enthusiastic and asked if he could meet with Derrel and me. I told him I would get in touch with Derrel and set up a time when he could fly to California for the meeting.

A few weeks later Derrel and I were in one of the inner offices of the William Morris Agency. I couldn't believe we actually had gotten this far down the road to success. The surroundings were not really impressive. The decor was probably at least ten years old. There were pictures of famous screen personalities on the walls, but no decorative frills, just plain desks with wooden chairs. A rug ran the entire length of the long hall and crept its way into the individual offices.

We were greeted by the receptionist. When we told her we were there to see Mr. Farnsworth, she said he would be down shortly. We waited about ten minutes, then noticed a gentleman wearing a dark business suit approaching the lobby from one of the adjoining hallways. He came into the room and spoke with the receptionist, who introduced us.

Farnsworth politely asked us to follow him to his office, and indicated that we would all wait there for two other associates who were joining the meeting. As we entered the office I noticed that the ambiance had improved. I was impressed and thought that perhaps we were dealing with one of the more favored executives. This gave me optimism about our upcoming meeting.

Two more executive types entered the room. Farnsworth made the introductions. One of the men, Jack, was responsible for placing scientific programming and the other, Glenn, was in charge of general productions. They explained to us that the agency was so large and covered such a wide segment of the entertainment industry that their functions were divided up.

Derrel and I made our presentation, telling the story of our research. We explained the formation of F.I.R.S.T. and how we needed to fund our ongoing research. We told them we had enough material for a great television special. Everyone appeared to listen attentively.

Farnsworth was the first to comment. ''I think that we might have something here. I'm thinking about setting up an appointment with Lee and Joe over at Rhino Productions. I think this is just their kind of thing.''

He turned and looked at both Jack and Glenn, who were already nodding their heads. Jack was the first to speak.

''Yeah, I think that's a good idea, and maybe we should also call Art with DDIX.''

Farnsworth looked at Glenn and asked, ''What do you think?'' Glenn thought these were both good ideas. Then Farnsworth sat forward in his chair and in a low tone of voice said to us, ''I think you guys are right on the money. Let's go ahead and make the calls.''

With that, both Jack and Glenn told us how pleased they were to meet us, and left the room. Steven Farnsworth settled himself into the chair behind his rather

large executive desk and said, "Gentlemen, I'm going
to make a couple of quick calls and see if we can get
some opinions for you."

Derrel and I sat back in our seats and smiled at
each other.

Farnsworth spent about ten minutes on the phone,
and then leaned across his desk and said, "I have made
arrangements for you to talk with two producers who
are clients of ours. If you are satisfied with what they
come up with, we will meet again and talk about sign-
ing you up."

Derrel was the first to respond. "I would like to
know what your fees will be if we decide to go forward
with the deal," he said bluntly.

"We only charge six percent of the gross earnings
per deal. It's pretty standard in the business."

We turned and looked at each other. I thought that
was a fair fee, and could tell by the expression on Derrel's face that he felt the same way. With that we stood,
shook hands with our host, and told him we would be
back in touch.

About two weeks passed before we were able to get
together with the producers Farnsworth had recommended. The meetings were less than satisfactory. They
were interested in doing a deal but would allow us almost nothing in the way of profit. When we asked them
why we should proceed with an unprofitable deal, they
told us that exposure was what we needed and this
would eventually lead to some moneymaking deals for
our organization.

Derrel and I felt like we were definitely getting the
short end of the stick. We were beginning to understand

the situation: profit for the big boys but none for us. Derrel returned home, a bit discouraged.

It was sheer coincidence that I received a call from my friend Dr. Tal. He was interested in how things were going in general. I explained our situation and he told us that he still had many friends in the entertainment business and would be glad to help.

In just two hours I received a call back from him. Tuesday afternoon was satisfactory for all parties, and we were to meet at a small Italian restaurant in Encino. He gave me directions and told me the time was set for 2 P.M. He also advised me to bring along any visual materials I had available.

The lights were dim in the small restaurant, and the aroma of oregano filled the air. There were only about fifteen little round tables scattered about the room, covered in red tablecloths topped with white paper. I looked around and saw Dr. Tal sitting at a table in the corner. He raised his hand and gestured for me to come over. Two other gentlemen were with him.

I approached and was introduced to Larry and Bob. They were with a company called Apollo Productions. I knew enough about Dr. Tal's eating habits to know that he did not drink alcohol and loved bread. Therefore I was not surprised when a waiter came over with a large basket of Italian sourdough bread. Dr. Tal started in on it immediately. I asked if anyone intended to order other food, but just bread and drinks were the order of the day.

Dr. Tal broached the subject of our meeting. I described what Derrel and I had been doing and why we wanted to use the entertainment industry to get our message out. When I finished my presentation, Dr. Tal got right to the point.

"How much money would a project like this take and how much could we make up front?" he asked bluntly.

"Well, I think if we do this right we are talking about a five-hundred-thousand-dollar budget," Larry answered.

Bob nodded his head in agreement and asked, "How much did you consider your share would be?"

Tal looked Larry squarely in the eye and without hesitation said, "We need at least two hundred and fifty thousand up front. In addition we want at least ten points on the back end for home video and offshore distribution. If that can't be done, we'll go elsewhere. Did I tell you boys that Scott Alexander is interested in writing a screenplay and doing a major motion picture on this deal?"

I felt about two inches tall. The numbers these guys were throwing around were staggering. I thought, "Could this actually be money they're talking about?"

Before I had a chance to speak, Larry turned to me and said, "Doc, could you tell me about an abduction episode one of your patients was involved in?"

"Larry, I'll be more than happy to." I spent the next few minutes telling Patricia's story about the camping trip and the old iron bridge. Larry sat listening intently. I noticed that while he was wrapped up in the story, he was also doodling with a pencil on the white paper that covered the tablecloth. The more he became involved in the story, the faster he doodled. Finally, at the conclusion of the story, he slowed down and stopped altogether. I had told him about the light in the sky and how Patricia's husband thought it was a truck. I explained how she finally realized it was a flying craft, but at no time did I describe the craft itself.

Larry and Bob left to come up with a treatment for our story, and arrangements were made to meet again at their office. Dr. Tal and I accompanied them outside. I excused myself and ran back to the table where we had been sitting. None of the busboys had been there yet and everything was as we had left it. I rushed over to the side of the table where Larry had been sitting and peered at his drawing. I could not believe what I was seeing. The picture Larry had drawn was an exact replica of the craft that Patricia had seen. I grabbed the paper, tore off the section with the drawing on it, and stuffed it into my pocket. I didn't care if anyone was watching me. When I showed it to Dr. Tal later, he was in a total state of disbelief.

Several weeks went by before we heard from Larry again. He wanted to set up another meeting, this time at the Apollo offices. I called Derrel and he suggested we set it up as soon as possible, even though this meant he would have to make another trip to California when I told him about the drawing Larry had doodled. Derrel was convinced that Larry could be an abductee who was in a typical state of denial. If this was true, we didn't know if it would help or harm our project. I called Dr. Tal and Larry and arranged the meeting for the following week.

Larry's office was in a high-rise building on Ventura Boulevard, one of the main streets in the San Fernando Valley. Derrel and I arrived slightly early. Dr. Tal was to meet us there. We took the elevator to the ninth floor, and when the door opened found ourselves immediately in the offices of Apollo Productions; evidently their offices occupied the entire floor. We approached the pretty blonde receptionist and asked for Larry. She knew he was expecting us and asked us to have a seat. Within moments Larry came out. I introduced him to Derrel, then we followed him through a complex set of hallways. I noticed many small rooms enclosed with half walls of glass, all bustling with activity.

He led us into a room with a large wooden conference table in the center, surrounded by chairs. On the walls hung several blackboards with production schedules written on them. Framed posters of older motion picture titles decorated the walls, and loose papers covered with writing were stacked on the table. The decor did not appear to be new and that, in my opinion, was a big plus. I was coming to the conclusion that many producers have enough seed money to make a good impression on prospective clients but no real financial backing and basically no funds for production. They can walk into a multistoried office building, make a deal for a large office without paying any front money, sign a long lease, lease all the furniture, and bingo, they have an office in which to impress their unsuspecting victims. This was certainly not the case here. This office appeared to be a working facility and showed signs that it had been functional for some time.

Dr. Tal joined us within a few minutes, and then we were introduced to several other men who were part of the production staff. The last to arrive was Bob. We all took seats at the conference table. Larry began the meeting by reviewing the material we had discussed with him at our last meeting. Suddenly, his personality changed and he became almost antagonistic. He explained that the only way to properly document what we were doing was to do it all over again. The shocked expression on my face must have expressed my feelings, but Larry carried on with his suggestions. He told us that most of the surgery could have been faked and that if he were to proceed with this type of production, he would have to make sure that everything was absolutely aboveboard. He suggested performing another surgery with his crew there to film the entire procedure. In addition, he was going to install metal detectors outside the surgery room so he could prove to the world that the surgeons and other personnel could not sneak foreign objects into the surgery room and fake their extraction. As Larry continued, Derrel and I glanced uneasily first at each other, and then at Dr. Tal.

Dr. Tal was the first to speak. He completely changed the subject, going right to the matter of money.

"What kind of budget do you have in mind for this type of a deal?" he asked.

Larry looked up from the table with a detached expression and mumbled, "Well, Tal, in order to do this right, we're going to have to have a fairly decent-sized budget."

Dr. Tal immediately came back with, "Just how much do you think that would be?"

"Oh, I would say in the neighborhood of about seven hundred grand. Do you think that's accurate, Bob?" He turned to his associate.

Bob was in agreement, and brought the others into

the conversation. "Yeah, Tal, I think we can get this done for about that figure. We have to have a good product if we hope to sell it to the majors," Larry added.

Dr. Tal's large eyebrows were raised. He stared directly at Larry and asked, "How much would we get out of that figure?"

"Well, you'll probably be entitled to a licensing fee, and then we could give you a bit for other things such as talent and stills" was Larry's answer.

"Just what would that come to?"

"We could probably squeak out a few thousand bucks for you guys. You've got to realize that you'll be able to do a lot better on the next one, and maybe there could be some back-end spin-off—you know, maybe a home video or other TV series."

At that point, Dr. Tal, Derrel, and I were all thinking the same thing. We politely excused ourselves and thanked Larry and Bob for their time. We have not heard from them since.

It was shortly after this that we met Chris Wyatt, then a CBS contract producer. He was extremely interested in the field of ufology and had produced the best UFO documentary Derrel and I had ever seen. Chris took a look at our work and offered to try to get CBS to do an hour special. He made such a favorable impression on us that we signed a contract with him.

As it turned out, a year has passed since that contract was signed, and there is still no television production. Chris left CBS and struck out on his own. We still have faith he will be able to get the job done.

Meanwhile, Derrel and I have appeared on several

television programs since we decided to release our information to the general public, including the series *Strange Universe, Paranormal Borderline,* and *Hard Copy.* We have also been on both Warner Bros. and NBC News, and appeared on the Iranian Television Network with a six-part series of interviews. In addition, we have presented our material on more than 250 radio programs around the world. Some of these include *Jeff Rense's Sightings on the Radio, The Rob McConnell Show, Hieronymus & Co.,* and Whitley Strieber's *Dreamland.*

CONFIRMATION

I N late July of 1998, I received a telephone call from Whitley Strieber, who had become a good friend through the years.

"Roger, I have just arranged a television special, called *Confirmation,* that is going to be shown on NBC during prime time. It will be a two-hour program of the kind that has never been seen before, showing some of the best physical evidence in the field of ufology. Are you interested in participating?"

I stood there in a mild state of shock. I thought about the opportunities this might bring. Without hesitation I answered, "What part would I play?"

"Roger, I think I can talk the production company into footing the bill for an entire surgery. In addition, I believe we could get them to pay for the laboratory testing of the materials.''

"Are you telling me this seriously?'' I asked.

"Roger, I am dead serious.''

It took just a few seconds for me to recover my composure. "I would be happy to do it. When are they planning to start filming and when would they like the surgery to be done?''

His answer gave me a jolt.

"Well, Rog, they want to start shooting right away and would like you to do the surgery in the next week or so.''

"Whitley, do you realize what I have to do to get all this done on such short notice? I have to locate the right surgical candidate, make arrangements for rental of a surgical suite with the proper equipment, notify the surgical team, have the patient undergo a presurgical examination, have his labs done, arrange for the MUFON camera crew, and take care of lots of other details. How do you expect me to get all this done in such a short amount of time?''

"I know it's short notice, but I know you can do it,'' was his reply. "What I want you to do first is to prepare a budget and submit it to me so I can present it to the production company. How long would it take you to do that? Can you get it to me in the next few days?''

I told him I would try to get it to him as soon as possible.

In the office the next morning, I placed calls to the members of the surgical team I'd used before. I also called Dr. A. and told him about the plan. Next, I called some of the prospective surgical candidates. One of them was very responsive but lived in Ohio. He had an object that was radio-opaque on X ray and sometimes caused a spark when he talked on the telephone. He was quite willing and anxious to come to California to have the object surgically removed, but I would have to foot his bill for the airfare. In addition, it would take a top surgeon to perform the surgery, since it was difficult to determine whether the object was nearer to the inside or the outside of the jawbone.

Some of the other prospective patients could not be reached at all. Others promised to call me back. Two cases told me they had decided to wait awhile before having the surgery done. Another case was a registered nurse who lived in California and had a radio-opaque object in her foot. She did want to have the surgery done, but upon reviewing the X rays, I noted the object was deep in the bottom of the foot and knew this would be a difficult extraction. Our best possibility was still the jaw case.

With X rays in hand, I headed for my dental associate's office to get his opinion. His name is Dr. David Schoenbaum, and he's been a friend for over thirty-five years. He was aware of the implant surgeries and had consulted on some cases that involved the mouth. I showed him the films, and he told me he thought the job should be done by an oral surgeon. He gave me a business card with the name of a friend who he thought might be interested in doing the job. So I was off again to consult with yet another doctor.

I took the elevator to the third-floor office of an Encino medical building. The elevator door opened and I began looking for suite 343. I soon found myself standing before a door that read, "Dr. Abraham Schlesinger, Oral Surgery."

I went in and introduced myself to the receptionist, who seemed to know who I was. She told me the doctor would be right with me. Soon I heard my name called and saw the door to the inner office open. A small woman with long blonde hair beckoned for me to follow her. She led me to a consultation room and asked me to have a seat, adding that the doctor would be right in. Only a few minutes passed before the door opened and a tall gray-haired gentleman with dark-rimmed glasses entered the room.

"Hi, I'm Abe. David Schoenbaum said you would be coming by. Let's take a look at the X ray." I handed him the film and he thrust it up on the dental view-box.

"Wow, this is an interesting one," he offered.

I stood and watched as he carefully scrutinized the film. I thought for sure he was going to offer to become the surgeon who would extract this object, so I was a bit shocked when he told me that this was not a job for an oral surgeon because it required an approach from the outer surface of the face. I was also disappointed to find myself back at square one again. I thanked him for his opinion and returned to my office, which by now was full of patients.

The following day I called my friend the dentist and asked him if he knew a surgeon who specialized in maxillofacial surgery. He recommended a specialist in that field who just happened to be in my immediate area. I thanked him, made the call to the new surgeon, and set up a time for a consultation.

The next afternoon, Dr. Fine reviewed the X ray

and told me that he would be able to do the surgery. I sighed with relief. I told him some of the details and circumstances regarding the research project and explained that there wasn't a large amount of capital available. I asked him what his fee was going to be. He looked at me with a straight face and told me it would be $4,000. I thanked him very much for his time and headed back to my office. I knew at that point that the case was just not going to work. The budget would be too large and would never be approved.

The phone calls I had made earlier became productive. I was able to get a tentative agreement from my regular surgical team and other MUFON volunteers. I also notified Derrel, and he said he could make the date. My next stroke of luck came when I was able to nail down a surgical suite that had the necessary X-ray equipment. This facility would cost $2,500, which was reasonable, considering it came complete with equipment and staff. Also, we could use it for the entire afternoon.

A possible date of August 17 was set. I told them I would let them know definitely in a few days. Dr. A. couldn't make the date, but told me Dr. Mitter would be glad to do the job. He was the same surgeon who performed the procedures in May of 1996. So I had everything lined up except for the patient.

The following day I placed another call to Whitley and told him about the progress I was making. He said he would be coming to California soon and would be staying until the production was finished. I thought that was a great idea. Somehow, down deep inside, I knew my one remaining problem would be solved.

At about eleven o'clock the next morning, my secretary told me I had a telephone call from someone who said he'd talked with me at the last MUFON meeting. She told me his name was Paul. I don't usually take calls from strangers during professional hours, but for some reason, I decided to take this one.

Paul refreshed my memory immediately. I had met him at the last local MUFON meeting, where he'd told me he had a foreign object in his thumb and was concerned that it could possibly be an implant. I'd asked him how he knew the object was there and he'd told me he had seen it on an X ray. I'd asked him if he could obtain the film and he'd said, "Sure, I have it right here with me."

He'd produced the X ray of his left hand, and sure enough, there was a bright shiny metallic object visible in the area of his thumb. I had asked him to come to my office so that I could take another film and rule out the possibility of its being an artifact. He had agreed, but had not called me until this morning.

I couldn't believe this strange set of events. I thought, "Could this actually be the surgical candidate I'm looking for?" I asked him if he could come to my office ASAP, and he agreed.

The next day, Paul was right on time. I reviewed his original X ray and confirmed the presence of a foreign object for the second time. I immediately took another X ray and developed it. Sure enough, the film showed the object to still be present. I was now sure he was going to be our surgical candidate, and couldn't wait to tell Whitley the good news.

When the last patient of the day departed, I sat

down at the computer and began drawing up the budget Whitley had requested. As soon as I finished, I gave him a call. He had already moved to California, and his Texas number referred me to a number there. When I got him on the phone, I related the good news and at the same time told him I had just finished the budget. He told me to fax it right away; he would go over it and then submit it to the production company for their approval. He indicated he would get back to me shortly. I left the office that day with new hope.

I should have remembered my past experience with the television industry. One of the lessons I'd learned was that they never approved the initial budget. When I called Dr. A. and filled him on the latest events, he told me I should have consulted him before submitting it. He said they were going to tear the hell out of it and we would be left with expenses that could not be paid.

The next day at about 2 P.M., he came to my office and reviewed the already submitted budget. He thought the only thing we could do was try and trim some of the expenses in preparation for the forthcoming rejection. By doing this, we would be one step ahead of the game.

Dr. A. knew the same people I did in the medical profession. He spent the rest of the afternoon on the telephone discussing the surgery with everyone who was to be paid out of the budget. He made agreements with all the participants that if their fees were rejected they would accept what was offered. This was good news. I felt as if another giant weight had been lifted from my shoulders. Now all I had to do was wait for Whitley's phone call.

At about 7 P.M. Whitley called and told me he

would put me in touch with the producer and that I should discuss the budget directly with him. About an hour later I received a call from a gentleman who introduced himself as Star Price with Mark Wolper Productions, the company making *Confirmation* for NBC. He told me he was looking forward to filming the surgery. He also wanted me to provide additional material such as photos, videos, and past laboratory reports, as well as biographical information about me, Derrel, and the proposed patient.

I asked him if had gone over the budget. He told me he had begun chopping the amounts for each item, explaining that the budget for the show was not that big. I recognized this as the stock line given by every production company I had ever worked with in the industry. We went over it point by point and I gave him my opinion about what could be cut and what could not. Finally we reached an agreement that I could live with and I began to feel good about the entire deal.

Whitley had hoped my budget would be big enough to include a fee for myself, but it didn't end up that way. I also was not told the production company was going to produce a video which would use all of my material and that I wouldn't receive anything from all the money they made.

In thinking over this episode, I have come to the conclusion that I was fortunate that at least some of my goals were fulfilled. I was able to perform another surgery, and in addition, there was some scientific analysis of the object done.

Suddenly it was August 17, the day of the surgery. I

loaded my car with several pounds of releases, consents, and other medical forms and arrived at the surgical center early in order to get ready. We had not
invited as many witnesses to this event because of the
small size of the waiting area which would house the
closed-circuit TV.

The first to arrive were four members of the production crew, along with Star Price. He told me the
remaining crew would be there with more equipment in
just a few minutes and asked me to show him the facility so they could decide where to place the television
cameras. I gave him a tour and introduced him to the
staff. I also advised him that our own team would be
filming the procedure, strictly for the scientific record.
He was less than happy about that, and made me promise that the footage we shot would not be used for commercial purposes.

Soon other members of the surgical team began to
arrive. I was busy rushing from one spot to another,
trying desperately to coordinate everything. It wasn't
long before problems began to arise. First, we could not
get the TV in the viewing area to work, so another TV
had to be moved from upstairs. Then there were technical problems with the size of the cable fittings, and on
top of that, there were conflicts between the professional television crew and our MUFON volunteers as to
where our camera would be set up in the operating
room so it did not interfere with the professional cameras. Our camera was finally placed in a corner, high
above the operating area. The viewfinder could only be
reached by standing on a ladder.

Adding to the confusion were the rules of the surgical facility and their enforcement by the operating room nurse, who is essentially the boss of everything that happens in her domain. She held a meeting during which she informed all nonprofessional personnel about what they could and could not do in the operating room and told us the number of people who could actually be present during the surgery. There were only six available sets of X-ray protective gear available, and everyone in the room would be required to wear these special leaded garments. This presented another problem, because the production crew—which included the producer, Star—exceeded that number by two. Star made it clear that if he was not in the room, no filming would take place. The problem was solved when Mike Evans, our MUFON surgical nurse, suggested that Star be allowed to stand behind a lead screen, which was already in one corner of the operating room. The O.R. supervisor agreed and said that when the X ray equipment was in operation, he would have to go behind the screen. However, we still had the lighting technician and the sound man. It was decided that the lighting technician would leave the room after he made sure the lighting was satisfactory, and the sound man was banished to a small supply closet with a glass door so he could observe the events.

The level of activity began to increase with the arrival of the patient, and our guests, including Whitley; his wife, Anne; and other volunteers from MUFON. The most important person on the team yet to arrive was the surgeon.

I made my way to the dressing room and changed

into my surgical greens; most of the crew had already done this. Then I heard Dr. Mitter's voice in the hallway. I sighed with relief as another piece of the puzzle fell into place.

When I got to the operating room, everything seemed quite orderly and relaxed. Then something caught my eye. I was mortified to see that our MUFON camera had been moved and now was not pointing at anything in the surgical area. I almost lost my composure.

"Who moved our camera?" I shouted to anyone who happened to be listening.

The answer came quickly from Star, the producer. "Oh, we had to do that because we needed to put one of our own cameras up there."

I decided to give him a taste of his own medicine and said, "If that camera isn't covering the surgical area, there will be no surgery done here today."

With that, one of the technicians bolted over to the ladder by our camera, climbed it, and tried desperately to peer through the lens. He announced that everything looked okay. I wasn't convinced, so I decided to take a second look at the monitor that displayed the surgery to our witnesses. It seemed to be just fine, and I returned to the scrub area.

Before putting the six pounds of protective lead shielding on me, the sound man placed several lapel microphones just under the front of my collar so that they could record my narration. All the members of the team were now in position, and I took my place slightly behind Dr. Mitter. Derrel was off to the side, sitting near Paul's head.

Just as I was getting into position, I felt a hand on my arm and saw Star standing there, his outstretched bare hand holding the sleeve of my sterile gown. He muttered something, but before I could hear what he said, the feisty little operating room supervisor saw what was going on and shouted sharply, "You there! Get the hell away from Dr. Leir right now and go back to where you are supposed to be or you will have to leave the room immediately."

Poor Star finally realized what he had done and sheepishly moved back behind the lead screen without uttering a word. Next I heard Mike shout, "Someone get Dr. Leir a sterile cover sleeve."

The scrub nurse helped pull that sleeve over my gown sleeve and then helped me replace my surgical glove. Finally we were all set to go. Dr. Mitter picked up the scalpel and made his initial incision into Paul's thumb.

Time seemed to pass quickly after that. The X ray equipment allowed us to see the foreign object on the television screen as it was being removed. The surgeon penetrated the area with a sharp probe until he could see that it touched the foreign object. The patient was doing well. Finally, Dr. Mitter asked for a surgical clamp. He lowered it into the wound and announced to all concerned, "I've got it."

With that he produced an object and placed it on a gauze surgical sponge. Although Dr. Mitter certainly knows what our surgeries are about, he remains pretty much a nonbeliever. He looked at the object on the gauze square and said, "My, that is unusual, isn't it?"

This was my first glimpse of the implant. When I

looked at it, the hair began to rise under my surgical cap. I heard someone say, "What is it? You had better tell the audience—don't forget we're recording this."

"Yes," I stammered. "It seems as if we have another one of those small cantaloupe seed-shaped objects, which appears to be covered by this very dark, gray, dense, well-organized membrane. I am now taking a surgical blade and trying to cut it open. It seems as if it won't open. My word, it's just like the other ones."

The cameras began to close in, aiming at the specimen. Within a matter of minutes, the patient was taken to the recovery area. The production crew removed their equipment and the nursing personnel began to take away all the used linen and draping. At that point, Star gave me instructions about how they were going to set up the next scene. He told me I was to stay in the room, still dressed in my surgical garb, and that when I was given the signal I would walk through the surgical suite doors to where the pathologist would be waiting. I was then to hand the surgical specimen to the pathologist and have a short conversation about what he intended to do with it.

What I didn't know was that they were going to shut the air-conditioning off during the time I was waiting in the operating room. The temperature slowly began to rise. I was wearing all my surgical gear, as well as the six pounds of lead. The longer I stood there, the more uncomfortable I became. All this time I was holding the surgical specimen in my hand, still on the original white gauze sponge. I looked at the object, and was shocked to see that it appeared to be getting smaller.

Was the temperature getting to me and my vision somehow becoming distorted? I picked up a clean surgical
towel and wiped my eyes, then looked at the specimen
again and realized that the damn thing was indeed
shrinking right before my eyes. I thought, "What
should I do? If this continues, when I finally walk
through those doors, I'll have no specimen left to talk
about."

At that moment I heard the prearranged knock on
the surgical doors and I started my walk. Dr. Roscher,
the pathologist, was standing there waiting for me. I
knew I had to do something fast. The first words out of
my mouth were directed to the operating room nurse.
"Where are the tubes containing the serum solution
that was removed from the patient?" I asked sharply.

The answer came swiftly from Dr. Roscher. "They
are right here," he said with a thick European accent.

I reached over and snatched them out of his hand.
The operating room supervisor handed me a pair of
surgical forceps and I used them to take the object off
the surgical sponge and plunge it into the vial of waiting blood serum. I was not concerned with the smaller
soft tissue specimens left clinging to the gauze
square—my primary concern was the solid object we
had removed. I am sure no one knew how urgent this
was. I had to get that object safely into the vial of
serum. No one seemed to comment on the sudden
change of plans. The next thing I heard was "Cut!
That's a wrap."

Very few people know the full story of what went
on in that room during those minutes when I was standing there holding the surgical specimen. Fortunately,

when the object was put into the patient's own serum solution, it began to return to its normal size. Perhaps the membrane surrounding the object started to dehydrate, and when placed back in the solution rehydrated. Then again, there may be an explanation of a more complex nature. This is just one more mystery surrounding the issue of implants and their removal. By this time I had learned there is no substitute for experience. Unless a surgeon has had training in the removal of these objects, he or she should not attempt this type of operation.

POST-OP
C H A P T E R   F O U R T E E N

ONCE the taping was over, I began to relax. I made my
way to the viewing area to greet our guests. They all
seemed happy and in good spirits. When I asked
them what they thought about the procedure, I was
shocked to hear them tell me they could not see a thing.
The only thing that came into view was my back and
the backs of the other operating personnel. I instantly
knew what had happened. When the camera had been
refocused on the surgical area, no one had checked to
see whether any of the participants might be obscuring
the view. Suddenly I realized that I would have no
video record of the procedure.

My one hope was to find Mike Portanova, our official MUFON photographer and videographer. I rushed

back to the dressing area and located him, and told him about the problem.

"Don't worry, Rog," he consoled me. "While you folks were all involved with the surgery, I was walking around with my new handheld video camera, shooting everything there was to see." I sighed with relief.

The object was now in its container, in the possession of Star Price. A document stating that this was the implant we had just removed was signed by all the major participants, including Whitley. The next day the object was brought to my office, where, with the cameras rolling, I removed it from the container and placed it in a drying container. The new container was sealed with a special tape and the document passed among those participating for a second set of signatures. The specimen stayed in this container for twenty-four hours.

The following evening, we met again at my office to attempt to remove the membrane. With the cameras rolling, I carefully removed the object from the drying container and placed it on a glass slide. It appeared to be dry, as I had expected. I took the surgical blade and made a small incision on one end of the dark cantaloupe seed–shaped object. Much to my surprise, I was able to scrape a small hole into the membrane and then carefully peel away the outer soft tissue covering.

"Wow! Take a look at this, guys. I was able to get this thing off almost in one piece. I think I really have the hang of it now."

Everyone gathered around, trying to get the first glimpse of the material that had been contained inside the membrane. They saw a small dark metallic rod, measuring just a few millimeters in length and about the circumference of pencil lead. In addition, there was a small pile of what appeared to have once been living membrane. There was one large piece among the debris that reminded me of a shell that had just come off a peanut.

The cameras panned the pieces closely. I suggested we check the magnetic properties of the object, then picked up a small magnet and held it about four inches away. I was astounded to watch the metallic object begin to shiver. As I lowered the magnet, it began to move. At about two inches from the object I was able to use the magnet to actually turn the object in a 360-degree circle without physically touching it. We all agreed it must be highly magnetic and most probably contained iron. I suggested trying an additional procedure with the magnet. I turned the object around 180 degrees, stopped, withdrew the magnet, and approached with the magnet again. To our amazement, the object spun around 180 degrees to its original position.

We then placed the object in its final transport containers, sealed them with the special tape, initialed the tape, and made a statement before the cameras. In addition, the document was signed again. The soft tissue membrane stayed with me, and I turned it over to Dr. Roscher within a few days. The metallic object was to be conveyed to a laboratory in San Antonio. Whitley would be there in the lab with the production crew to witness the testing. They would get the results to me as soon as the testing was performed.

The first results were reported to me over the telephone by Whitley. He told me that the results were earthshaking. He explained that the test they chose to perform was called an X ray diffraction exam, in which, after it is prepared and mounted, the specimen is exposed to a controlled X ray beam and a segment of sensitive film. The results after exposure to the X rays are circles that appear on the film when it is developed. This data is fed into a computer database, and the computer then graphically portrays the elements it has identified. This demonstrates what materials make up the sample.

Whitley explained to me that in this case the standard procedures were performed, and the result was a graphically recorded peak that could not be identified by the computer, although it contained a database of over sixty-five thousand known elements and compounds. The scientists were so amazed, they began to doubt the veracity of the test, so like good scientists everywhere, they decided to repeat it. Not only was it repeated once, but they evidently did the test fourteen times. They also tested the machine to make sure it was working correctly. Each time they tested the implant, the result was the same: the substance it was made out of could not be identified.

What did this mean? By all logic it should have meant that the sample had no organized atomic structure, which would have indicated that it was just an amorphous mass. But how could this be, when it demonstrated all the properties of magnetic iron?

The next report to arrive was the pathology results. Dr. Roscher's report is probably one of the most complete and detailed pathology summaries written so far in the study. His background in the field of pathology is very extensive and his reputation is known worldwide.

In the soft tissue material that was removed adjacent to the object, he found no overt inflammatory reaction or foreign body reaction. The membranes showed areas of brown hemosiderinlike deposits, sporadic nerve trunks, and fibro-tendinous tissue.

This analysis repeats the findings of our previous reports and again raises questions pertaining to biology and pathology that still need to be explained. The most significant finding is that there is absolutely *no reaction* by the body to the foreign object. This is still impossible, by our standards of knowledge. The other significant reproduced finding was that the membrane contained protein elements of blood and hemosiderin. The report contains the following statement:

"The absence of an overt inflammatory process and the absence of foreign body reaction is indicative that the structure removed from the hand is of unknown nature at this time." In addition, it goes on to say, "The absence of a reactive inflammatory process, and/or a foreign body, indicates that the structure must be rather inert not to elicit a foreign body response."

So here we have corroboration of previous findings. More research will have to be performed to determine exactly what this all means.

I have followed this patient as well as the others since the surgeries were performed and am happy to say they all healed well and have no unpleasant side effects.

One of the major postoperative developments has been my separation from Derrel Sims as a working partner. There comes a time when differences of opinion begin to impede the progress of any partnership. This is what happened to us.

I have gone on to form a new organization, Aliens and the Scalpel Research, or more simply A & S Research. The purpose of my new group is to use all the scientific methods available to better understand the abduction phenomenon and to try and help abductees in very concrete ways. Information about our work can be found on the Alienscalpel.com website.

Financial concerns are still a major problem. It continues to cost many dollars to do this research, and as a nonprofit trust we are constantly looking for new sources of funding. I feel it is safe to assume I will always have to make an effort to get the funds necessary to do this work.

When I began my research into the abduction phenomenon, I was innocent and naive, and, after all my years in medical practice, thought I had become accustomed to stress. I had no idea what I was getting into, stepping into a field filled with not only strange events but, in some cases, strange people as well. Perhaps this is due to frustration over the lack of cooperation from established scientific organizations. It may also have much to do with the lack of funding to accomplish goals. In addition, there is frustrating government disinformation as well as the media's attitude that anyone involved in this field must be some sort of nutcase.

After I published a small, early edition of this book, I received notice from the California Medical Board that a patient had registered a complaint against me. I went on to learn that this complaint had been filed almost ten years ago and the board had sat on it for all those years without taking any action. In addition, I

discovered that the patient had passed away from unrelated causes some five years ago, the hospital involved
had closed its doors about seven years ago, and all the
records were unavailable. I was going to have a hard
time defending myself.

This reminded me of what happened to Dr. John
Mack, the Harvard psychiatrist who published a book
saying that he believed abductees were telling the truth
about their experiences. A group of fellow professors
tried to end his tenure and have his license revoked, for
what they said were "unrelated reasons." I suspected I
was now the object of a similar type of "witch hunt."

These events, coupled with my separation from
Derrel, began to take a physical toll. I had a heart attack
in June of 1999.

In 1985 I had undergone a six-vessel cardiac by-
pass procedure. This served me well for fourteen years.
Now my physicians have advised me that no further
surgery can be performed. I continue to follow a medical regimen, and have started taking natural products
designed to improve cardiac function.

My troubles are far from over. I still have to deal
with the medical board, support a full-time practice,
carry on my research and writing, and repay the huge
medical bills I ran up during my heart attack. I want to
thank my close friends who came through for me, for
their prayers and financial support.

It is my sincere hope that the research in this field
will continue, with more individuals in hard science
taking part in the adventure. I will be there as long as I
can, donating my time and energy to help the many

victims of alien abduction. The best help we can give them is to assist in providing proof that their experiences are *real*. Locating their implants and removing them, and eventually discovering their function, is one important way to do this.

A P P E N D I X
Scientific Studies

The implants have received extensive scientific analysis, both for their biological properties and their metallic content.

BIOLOGICAL ANALYSIS

Biopsy and pathology reports have been obtained through conventional sources. These confirm that the implants are encased in a membrane made of, among other things, skin. But skin cannot be grown by the body in deep muscle tissue, where these membranes have been found. This is because the body does not possess the genetic coding to grow skin anywhere except on the surface. And yet the tissue is there. It has the effect of minimizing the rejection response, thus making it possible for the implants to remain in the body for years.

It is unfortunate that this unique material continues to be ignored by science in deference to the fiction that the implants are explainable in some ordinary manner. As a result, the chance of understanding how to develop membranes that would prevent the rejection of medical implants is being lost.

METALLURGY

The metal fragments found encased in the membranes were sent for analysis to New Mexico Tech by the National Institute for Discovery Science. The laboratory was not given any information about the origin of the samples and came to the conclusion that they were probably meteoric without knowing that they had been surgically removed from people believing themselves to have been implanted by aliens.

Samples of the various scientific reports are included in this appendix. Further information on the samples and ongoing studies can be obtained from the websites of the National Institute for Discovery Science (*www.accessnv.com/nids*) and Dr. Roger Leir (www.alienscalpel.com).

MEDICAL LABORATORY _____, INC.   **Biopsy Report**

California

NAME: Age: 52   Sex: F
Roger Leir, DPM   95-MN-12443   Rec'd: 10-06-95
Thousand Oaks, CA

SPECIMEN: LESION, FOOT

GROSS: The specimen consists of a 0.3 x 0.1 x 0.1 cm irregular piece of dull,
brown tissue along with a similar piece half that size. Both parts are submitted.
/eb

MICROSCOPIC: Sections show a curved piece of proteinaceous, eosinophilic coagulum
admixed with keratinous material and coarse, amber brown granules. No malignancy is
identified. An iron stain confirms the presence of hemosiderin.

DIAGNOSIS: LESION, FOOT: KERATIN, PROTEINACEOUS COAGULUM
                 AND HEMOSIDERIN.

                                          _____, M.D.
/eb
File
10/10/95

MEDICAL LABORATORY , INC. **Biopsy Report**

California

NAME: Age: 47   Sex: M
Roger Leir, DPM   95-MN-12444   Rec'd: 10-06-95
Thousand Oaks, CA

SPECIMEN: LESION, LEFT HAND

GROSS: The specimen consists of a 0.3 x 0.2 x 0.1 cm brown and tan shred of
tissue. The specimen is fully embedded.
/eb

MICROSCOPIC: Sections show a strip of dense, eosinophilic fibrillary material
containing rare portions of superficial, degenerated epidermis and scattered, amber brown
pigment granules. No malignancy is identified. An iron stain confirms the presence of
hemosiderin.

DIAGNOSIS: LESION, LEFT HAND: PROTEINACEOUS COAGULUM
                    SUPERFICIAL DEGENERATED EPIDERMIS AND
                    HEMOSIDERIN.

_____

, M.D.

/eb
File
10/10/95

MEDICAL LABORATORY , INC. **Biopsy Report**

California

NAME: Age: 52   Sex: F
Dr. Roger K. Leir   95-MN-10911   Rec'd: 8-22-95
Thousand Oaks, CA

SPECIMEN: LESION, FOOT

GROSS: The specimen consists of eight tan irregular tan fragments measuring up
to 0.7 cm. The specimen is fully embedded.
/c

MICROSCOPIC: Fragments of fibroconnective tissue and fat demonstrating peripheral
nerve and pressure receptors. The peripheral nerve segments show perineural fibrosis. There
are included fragments of hyperkeratotic stratum corneum. No foreign material is identified
by either plain or polarized light, except for a few small polarizing fragments with the
appearance of cotton fibers. No inflammation is present, and there are no atypical cell
changes.

DIAGNOSIS: 1)   PERIPHERAL NERVES WITH MILD PERINEURAL
                     FIBROSIS.
                2)   FAT, FIBROCONNECTIVE TISSUE, AND SKIN - NO
                     DIAGNOSTIC CHANGES (TISSUE FROM FOOT
                     ADJACENT TO FOREIGN BODY).

_____
                                                    , M.D.
/cb
File
MN
08/23/95

LABORATORY REPORT

ROGER LEIR, D.P.M.   001.004

			ACQUISITION NUMBER		ROOM NUMBER	
			DATE COLLECTED	ORDER TYPE		OTHER IDS/LOCATION
			05/18/96			
PATIENT NAME		AGE/SEX	ACCT NO. / CHART ID #	PHYSICIAN	RECEIVED	REPORTED	SPEC. NO.
		F			05/24	05/31/96	8638952

***** P A T H O L O G Y   R E P O R T *****

CLINICAL IMPRESSION AND/OR HISTORY:

　CLINICAL INFORMATION NOT PROVIDED

BIOPSY SITE AND/OR SPECIMEN:

　LEFT CALF

GROSS DESCRIPTION:

　　THE SPECIMEN CONSISTS OF A WHITE TAN SKIN FRAGMENT, MEASURING
　0.7 X 0.5 X 0.2 CM. THE SPECIMEN IS SECTIONED INTO THREE
　PIECES AND SUBMITTED IN A SINGLE CASSETTE.

MICROSCOPIC DIAGNOSIS:

　SKIN, LEFT CALF, SHAVE BIOPSY:
　-FIBROSIS WITH SOLAR ELASTOSIS, NO EVIDENCE OF MALIGNANCY

COMMENT:

　SUPPLEMENTARY REPORT: 06/07/96
　THERE IS NO EVIDENCE OF INCREASED NERVE BUNDLES OR SIGNIFICANT
　CHRONIC INFLAMMATION. NO OTHER HISTOLOGIC EVIDENCE OF FOREIGN
　BODY REACTION IS SEEN.

　　　　　　　　　　　　　　　　　　　　　　　 , M.D.
　　　　　　　　　　　　　　　　　　SIGNATURE ON FILE
encl:(1) SLIDE/S

Physician service line, Tissue Pathology: (800)
Patient: Please consult your physician.
PAGE 1: END OF REPORT! FOR : 8638952

LABORATORIES   C4499 218
              BLVD.   ROGER K. LEIR, D.P.M.
    CA
(818)

                                                            PAGE: 1

                          NS   F   4499   LEIR   H0953120
NOT STATED
DOB: 06/07/96   06/18/96   FINAL
RM: NOT STATED
                                      SLIDE NUMBER: N9606592

BIOPSY

SPECIMEN SOURCE/SITE   LEFT CALF

GROSS DESCRIPTION
   Received in formalin is a single tissue fragment measuring 0.5 x 0.3 x
   0.2 cm. This specimen is bisected, totally processed. One cassette.

      :scr

MICROSCOPIC DESCRIPTION
   Sections show benign skin with minimal perivascular lymphohistiocytic
   chronic inflammation and mild orthokeratosis. No significant
   pathologic changes are identified.

DIAGNOSIS
   LEFT CALF: BENIGN SKIN WITH MINIMAL PERIVASCULAR LYMPHOHISTIOCYTIC
   CHRONIC INFLAMMATION AND MILD ORTHOKERATOSIS.

PATHOLOGIST   , M.D., 06/14/96   (:jpp)

                       *** FINAL REPORT ***

New Mexico Tech

Letter of Opinion (Samples T1,2 and T3)

The first theory on the origin of these samples was initiated due to the relatively high hardness value obtained for the iron core of sample T1,2. It is well known that very hard iron alloys can be found naturally in meteorite samples. In fact, several characteristics of the specimens are similar to certain meteorite-type materials. Meteorites can be a complex combination of many different elements (see for example, McSween, 1987). This is the case particularly for sample T3, which contains at the very least 11 elements: Na, Al, Si, P, Cl, Ca, Fe, Ni, Cu, Mo & Sn. Typical of iron and stony-iron meteorites is the classic "Widmanstatten structure", consisting of lamellae (plate or needle-shaped crystals) of kamacite (*alpha*-iron) and/or taenite (*gamma*-iron), formed during the slow cooling of meteoroids [McSween, 1987; Budka et al., 1996]. Interspersed with the metal grains are other minerals rich in iron and/or nickel such as troilite, FeS, and schreibersite, (Fe,Ni)_3P. Based on my examination, the samples in question could possibly fit into this framework. Elemental analysis done by X-ray Energy Dispersive Spectroscopy (EDS) indicated iron and phosphorus as major constituents of the cladding material surrounding the iron core. The (EDS) patterns resemble those recently reported for iron dendrites found in pockets and veins of the Yanzhuang II6 meteorite [Brooks, et. al., 1995]. In addition, I identified a calcium phosphate mineral as a possible phase within the cladding of both samples. Interestingly, chlorapatite, $Ca_5(PO_4)_3Cl$ is among the more common meteorite minerals [Wasson, 1974]. This would account for the presence of a substantial amount of calcium and smaller amount of chlorine detected. A problem with this theory, however, is that no nickel was detected in T1,2 and only a minute amount in T3. It has been stated that "most meteorites contain between 6 and 10 percent nickel"....and "no iron meteorites contain less than five percent nickel" [McSween, 1987]. This may not be a problem after all, since the specimens could be just a small fragment of a larger meteorite body.

An altogether different hypothesis can be formulated based on the fact that these specimens were extracted from an human body. An iron sliver, embedded in human tissue could possibly cause a calcification reaction. This would explain the presence of calcium and phosphorous on the surface of the samples. Chlorapatite and other calcium phosphate minerals are the major component of hard tissue (bones, teeth) along with collagen. In fact, calcium phosphate-based ceramics have been used in medicine and dentistry for nearly 20 years due to their bioactive nature [Hench, 1993]. In light of this, even if the cladding was not formed inside the body, but rather entered the tissue in its entirety as a sliver from a stone, it is not surprising that the body had no adverse reaction to the foreign object.

It must be stressed, these are only theories as to the origin of the specimens in question based on preliminary data and information. More in-depth studies would be required to prove either one.

August 28, 1996

Environmental Health & Safety Specialist
Organic Chemist

Let's 'talk' about the possibilities...

Sample #RR3, by the Nickel, Zinc and Silver ratios, may have extraterrestrial origins, but the Tungsten and Thallium ratios don't show any anomalies.

Sample #007KT, by the Nickel, Zinc, Ruthenium, Samarium, Europium and Tungsten ratios, may have extraterrestrial origins, but the Magnesium ratios (which have the largest signals) are consistent with earth origin.

Sample #AL101 shows some anomalies, for example with Germanium and Tungsten, but others are inconclusive. I doubt that there is any extraterrestrial origin with this material.

Sample #SM/DS shows the earth origin ratios for Magnesium and Zinc. Within some small variation, the other elements which have multiple isotopes detected appear to be close to the earthly natural abundance ratios.

In Conclusion:

It is possible, but not conclusively proven, that both the RR3 and 007KT samples show some isotopic ratios consistent with an extraterrestrial origin. More tests with a larger sample size would be required to know for sure. These samples were run without any idea of the chemical constitution of three of the samples. With the 'enclosed' data it would be possible to tailor subsequent tests to take advantage of this information.

Sincerely,

ABOUT THE AUTHOR

Dr. Roger K. Leir, podiatric physician and surgeon, received his Bachelor of Science degree in 1961 and his Doctor of Podiatric Medicine in 1964. He has been in private practice since that time. Dr. Leir has participated in research that included the study of regenerating tendons and the use of artificial substances in foot reconstruction procedures. Dr. Leir's interest in ufology began at age twelve when his father read the newspaper headline about the Roswell incident to his mother. In the late 1980s he became a member of MUFON, the Mutual UFO Network, and eventually functioned as an investigative reporter for the Ventura–Santa Barbara chapter's monthly periodical, *The Vortex,* which led to his later association with abduction researcher Derrel Sims. Dr. Leir has presented his research all over the world and was one of seven American researchers to be keynote speakers for the opening of the UFO-Aerospace Museum in Japan. Dr. Leir's implant studies were featured in Whitley Strieber's book *Confirmation* and his February 1999 NBC special of the same name. He is a regular guest on Strieber's radio program *Dreamland* and has appeared on radio and television programs around the world, including *Jeff Rense's Sightings on the Radio* and *Alt Bell's Coast to Coast.*